Jon Knaupp

# PLANE TRIGONOMETRY

# PLANE TRIGONOMETRY

Revised Edition

FRANK A. RICKEY, *Louisiana State University*
J. P. COLE, *Louisiana State University*

HOLT, RINEHART AND WINSTON, INC.
NEW YORK . CHICAGO . SAN FRANCISCO
TORONTO . LONDON

Library of Congress Catalog Card Number: 64–13403

**27446–0414**

Printed in the United States of America

# PREFACE

This volume, a revision of the authors' *Plane Trigonometry* (1958), is a continuation of the effort to give a brief, analytical presentation of the portions of trigonometry needed today. The acceptance given the first edition encouraged the authors to undertake a revision which provides new and more extensive sets of exercises, which attempts to improve the clarity and correctness of statements, but which does not make radical changes in the approach of the first edition.

Although the subject of triangle solution and its applications is given attention, the major emphasis is frankly on the relational aspects of trigonometry that give the subject its real importance. In choosing both subject matter and exercise material, effort has been made to contribute to the mathematical development of the student. Many of the exercises offer techniques and concepts of direct aid in the later study of mathematics and science.

Through the use of analytic methods, the student is encouraged to assume responsibility for his own progress. An abundance of drill exercises of a necessarily routine nature are provided, but a large number of exercises requiring varying degrees of thoughtfulness are included. An occasional "mettle tester" will challenge the ablest student.

Feeling that a text in trigonometry suitable for a three-semester-hour course is not the place to attempt a treatment of the foundations of algebra and geometry, the authors have assumed a knowledge of the real number system and of Euclidean plane geometry, reviewing, without proof, such concepts and results of these subjects as are needed. The notion of sets is introduced, not as "window dressing," but as an aid in a concise discussion of functions, graphs, and so forth.

Features of particular note are:

A chapter on preparations, providing the background for the analytical methods of the book.

The inclusion of general proofs throughout the book.

Stress on use of general results wherever this can replace memorization of special formulas. For example, use of the addition formulas, rather than numerous reduction formulas.

The early introduction of graphs familiarizing the student with the manner of variation of the trigonometric functions, including their periodic character.

Carefully developed exercises, including a miscellaneous set at the end of each chapter.

Graphical visualization of the roots of trigonometric equations, such as $\sin x = \frac{1}{2}$.

Rapid sketching of graphs in both rectangular and polar coordinates.

Adoption of S.M.S.G.-type notation where possible without sacrificing simplicity in trigonometric expressions.

A thorough treatment of complex numbers.

Appendix sections on approximate values of measures and on common logarithms.

Flexible arrangement of material. For example, teachers preferring to treat the subject of trigonometric identities in separate segments will find that the first part of Chapter 6, dealing with the "eight fundamental identities" can be used at any desired point after completing Chapter 3.

The authors are grateful for valuable criticisms and suggestions made by fellow teachers of mathematics at Louisiana State University and elsewhere.

*Baton Rouge, Louisiana*
*February, 1964*

**J. P. C., F. A. R.**

# CONTENTS

# PLANE TRIGONOMETRY

# CHAPTER 1. PREPARATION for the STUDY of TRIGONOMETRY

## 1.1 INTRODUCTION

The science of trigonometry (from Greek words meaning *triangle-measure*) originally developed as a means of computing unknown sides and angles of triangles and related figures. Computational trigonometry is still an indispensable tool of the surveyor and the navigator. But, as science has developed, its concepts have been broadened and generalized until today the language and methods of trigonometric analysis are fundamental in nearly every branch of science and engineering. Some of the ways in which particular parts of trigonometry are useful in other fields will be pointed out as the occasion arises.

We begin with an informal review of several important topics of plane geometry and algebra, upon which our study of trigonometry will be based.

## 1.2 SETS

Much of mathematics deals with sets—sets of numbers, sets of points, sets of circles, sets of rules, and so forth. The word *set* is taken to mean a collection of objects each of which is called an element of the set. We assume that there is some way of deciding whether or not a particular object is a member of a given set.

Sets are usually represented by capital letters, and elements of sets, by lower case letters. The symbol $\in$ is mathematical shorthand for the phrase "is an element of." Thus, in dealing with sets, the expression

$$x \in B$$

states that $x$ is an element of the set $B$.

Two sets are *equal* if they have exactly the same elements. If every element of a set $A$ is also an element of set $B$, $A$ is said to be a *subset* of $B$. The shorthand for this is: $A \subset B$. The set with no elements is called the empty set and is identified by the symbol $\phi$. Its role in the mathematics of sets is analagous to that of the number zero in arithmetic.

A set is usually identified either (a) by listing its elements within brackets, or (b) by stating a rule for determining the elements of the set. Examples of (a) are: $\{a, e, i, o, u\}$, the set of English vowels, and $\{1, 3, 5, 7, \ldots\}$, the set of all positive, odd integers. (The three dots are shorthand for "and so on.") An example of set identification method (b) is:

{every $x$ such that $x$ is an integer and $x$ is between 1 and 7}, which is, of course, the set $\{2, 3, 4, 5, 6\}$. In this type of set identification, the phrase "such that" is usually represented by the vertical dash symbol |, so that this set is represented by the expression

$$\{x \mid x \text{ is an integer and } x \text{ is between 1 and 7.}\}$$

Set $C$ is called the *union of sets A and B* (written $A \cup B$) if every element of $C$ is also an element of $A$ or of $B$. Here, the word *or* is supposed to include the possibility that the element may be an element of *both* $A$ and $B$. (In any case where *both* is not to be included, we are expected to say so.) Thus if $A = \{a, b, d, h, k, p\}$ and $B = \{b, e, h, p, r, s\}$, it follows that

$$A \cup B = \{a, b, d, e, h, k, p, r, s\}$$

Set $D$ is called the *intersection of sets A and B* (written $A \cap B$) if every element of $D$ is an element of *both* $A$ and $B$. Thus if $A$ and $B$ are the sets given in the previous paragraph,

$$A \cap B = \{b, h, p\}$$

One more bit of symbolism: a slash line / is used to denote the *negation* of many symbols. Thus $\neq$ means "is not equal to," $\notin$ means "is not an element of," and $\not\subset$ means "is not a subset of."

## EXERCISES

**1.** The blank space in each of the parts below may be appropriately filled by one or more of the symbols $=, \neq, \in, \notin, \subset, \not\subset, \cup, \cap$. State which symbols may be used for each blank.

**(a)** $\{1, 2\}$ — $\{2, 1\}$
**(b)** $\{a, b, d\}$ — $\{a, b, c, d\}$
**(c)** $a$ — $\{a, b, c\}$
**(d)** $\{a\}$ — $\{a, b, c\}$
**(e)** $2$ — $\{1, 3, 5, 7\}$
**(f)** $\{2\}$ — $\{1, 3, 5, 7\}$

**2.** List five members of each of the sets "built" below:

(**a**) $\{a \mid a = 3n + 2, n$ being any positive integer.$\}$
(**b**) $\{x \mid x$ is an English vowel.$\}$
(**c**) $\{r \mid r = 2^k, k$ being a positive integer.$\}$
(**d**) $\{p \mid p$ is a prime integer (divisible only by itself and one).$\}$

**3.** If $A$ is the set of odd integers and $B$ is the set of even integers, what is $A \cup B$? $A \cap B$?
**4.** If $A = \{x \mid x$ is an integer and 3 divides $x$ exactly$\}$ and $B = \{y \mid y$ is an integer and $y$ is even$\}$, identify the sets $A \cup B$ and $A \cap B$.
**5.** If $A$ is a set, is it true that $A \subset A$? Justify your answer.
**6.** State which of the following statements are true and which are false. In each case the capital letters refer to sets.

(**a**) If $A = B$ and $B = C$, then $A = C$
(**b**) $A \cup A = A$
(**c**) $A \cap A = \phi$
(**d**) $A \cap B \subset B$
(**e**) $A \cup B \subset A$
(**f**) $B \subset A \cup B$
(**g**) $A \cap B \subset A \cup B$

**7.** On accident-report maps, a street is represented by two parallel lines. Think of the street as the set of points between and on the boundary of the street. Show on a drawing (by shading) the set of points representing the *intersection* of the sets representing two crossing streets. How does this compare with the usual meaning of "street intersection"?

## 1.3 SETS OF POINTS ON A LINE. COORDINATES

Geometric figures are sets of points. In particular, a line (meaning a *straight* line) is a set of points. We assume that the reader is familiar with the elementary properties of lines* of which (1) and (2) below are examples.

(1) Two distinct points are on (are elements of) exactly one line.
(2) To each pair of points of a line is assigned a number called the distance between the points.

If $A$ and $B$ are two (distinct) points of a line $L$, there are several subsets of the points of $L$ that are of importance. These are:

---

* See any good high school geometry text. The notations given in this section will be found in the geometry material prepared by the School Mathematics Study Group.

(1) The line itself, symbolized as $\overleftrightarrow{AB}$. The double arrow is supposed to represent the infinite extent of the line in both directions.
(2) The line segment joining $A$ to $B$, written $\overline{AB}$. In terms of sets, $\overline{AB} = \{A, B,$ and all points of $L$ that are between $A$ and $B\}$.
(3) The ray $AB$, indicated by $\overrightarrow{AB}$, which is the half line with initial point $A$ that contains $B$. As sets,

$$\overrightarrow{AB} = \overline{AB} \cup \{\text{all points } C \text{ of } L \text{ such that } B \text{ is between } A \text{ and } C\}.$$

Figure 1.1 illustrates these three sets. Generally, arrows will be used on *drawings* of lines only when it is desired to emphasize the infinite extent. However, arrows will always be used in writing the symbols.

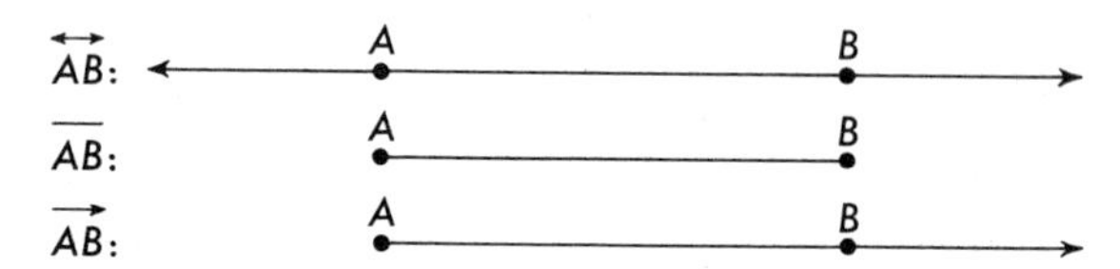


Fig. 1.1 Line $AB$, segment $AB$, and ray $AB$, that is, $\overleftrightarrow{AB}$, $\overline{AB}$, and $\overrightarrow{AB}$

The *length* of $\overline{AB}$ is represented by the unmarked symbol $AB$. It is the number of units of distance between $A$ and $B$.

If is useful to distinguish between the two directions along a line by regarding one direction as positive and the opposite as negative. If the line is horizontal with reference to the observer, the right-hand direction is usually taken as positive. If the line is not horizontal, the upward direction is usually regarded as positive. Figure 1.2 shows a horizontal line upon which a number

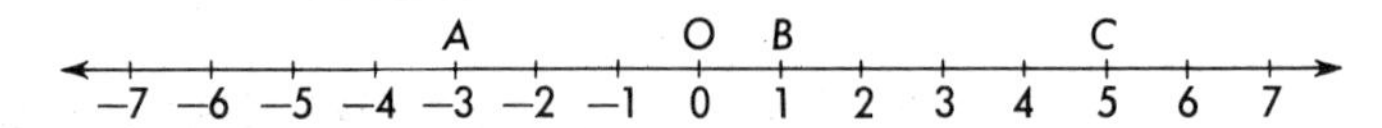


Fig. 1.2 A line with coordinates

scale has been set up so that each point on the line can be identified by a signed number that gives its distance and direction from an arbitrarily chosen *origin*, the point $O$ of the line.

The number that identifies a given point on such a scaled line is called the *line coordinate*, or simply the *coordinate*, of the point. To every point on the line there corresponds a number that is its coordinate. The endless set of numbers that thus corresponds to the set of all points on the line is known as the set of real numbers. The numbers $-1 + \sqrt{2}$, $\sqrt{5}$, $7/8$, and their negatives, are examples of real numbers other than the integers indicated on the scale. Unless otherwise stated, the word *number* will mean a *real number* in this book.

There is a general formula for finding the distance between two points on such a scaled line, if the coordinates of the points are given. If $x_1$ is the coordinate of the point $P_1$ and $x_2$ that of $P_2$, then*

$$P_1P_2 = |x_2 - x_1| \tag{1.1}$$

Examples, using Figure 1.2 are:

$$AB = |1 - (-3)|$$
$$= 4$$
$$BA = |-3 - 1|$$
$$= |-4| = 4$$
$$AC = |5 - (-3)|$$
$$= 8$$

and

$$CB = |1 - 5|$$
$$= |-4| = 4$$

The student can satisfy himself that this formula is true for all pairs of points on the line. Note that the relation $P_1P_2 = P_2P_1$ is satisfied by this formula.

## 1.4 ORDER RELATIONS

In assigning numbers to the points of a line, we used the assumption from algebra that the real numbers can be divided into three sets: $\{0\}$, {positive numbers}, {negative numbers}. Furthermore these sets are mutually exclusive so that their intersection is the empty set. This assumption is actually based on the so-called order postulate for real numbers, which may be stated as follows:

Given any (real) numbers $a$ and $b$, exactly one of the following is true:

(1) $a$ is equal to $b$
(2) $a$ is greater than $b$
(3) $a$ is less than $b$

In symbols,

(1) $a = b$
(2) $a > b$
(3) $a < b$

Taking $b$ to be zero, gives three possibilities for each number $a$:

(1) $a = 0$
(2) $a > 0$, in which case $a$ is called a *positive* number
(3) $a < 0$, in which case $a$ is called a *negative* number

---

* If $x$ is a number, "$|x|$," read "absolute value of $x$," means the positive number corresponding to $x$. That is $|x| = x$ if $x$ is positive or zero and $|x| = -x$ if $x$ is negative.

It is shown in algebra that the statement

$$a > b$$

is equivalent to the statement

$$a - b > 0$$

Likewise

$$a < b$$

and

$$a - b < 0$$

are equivalent statements.

Thus we say that $9 > 4$, since $9 - 4 = 5$ and $5 > 0$.

Other examples are

$$8 < 11, \text{ since } 8 - 11 = -3 \text{ and } -3 < 0$$
$$-5 < 2, \text{ since } (-5) - 2 = -7 < 0$$
$$-3 > -100, \text{ since } (-3) - (-100) = 97 > 0$$

Let us now make an observation about Formula (1.1), that states that if $x_1$ and $x_2$ are the respective coordinates of the points $P_1$ and $P_2$ on a scaled line, then

$$P_1P_2 = |x_2 - x_1|$$

We note that the number $x_2 - x_1$ (without the absolute value bars) may be zero, positive or negative. Of course, these possibilities correspond to the cases, $x_2 = x_1$, $x_2 > x_1$, or $x_2 < x_1$, respectively. In turn, these correspond to the cases (as related to the points) $P_2 = P_1$, $P_2$ is to the right of $P_1$ (in the positive direction from it), or $P_2$ is to the left of $P_1$, respectively. Thus we see that the number $x_2 - x_1$ may be used to determine both the distance $P_1P_2$ and the direction of $P_2$ from $P_1$ if they are different points. For example, if $P_1$ has coordinate $-9$ and $P_2$ has coordinate 5, then $x_2 - x_1 = 5 - (-9) = 14 > 0$. Hence $P_2$ is in the positive direction from $P_1$. Furthermore $P_1P_2 = |14| = 14$.

## 1.5 VARIABLE ON A SET. NUMBER INTERVALS

A symbol for which any element of a given set may be substituted is called a *variable on the set*. In Section 1.2, the set $\{2, 3, 4, 5, 6\}$ was described as $\{x \mid x \text{ is an integer and } x \text{ is between 1 and 7}\}$. The symbol $x$ is in this case a variable on the set $\{2, 3, 4, 5, 6\}$.

As another example of a variable on a set, consider the statement "$t$ is any number between 0 and 1." Here $t$ is a variable on the set of numbers from 0 to 1, not inclusive. A concise way to state this is:

$$t \text{ is a number and } 0 < t < 1$$

(Stating that $t$ is a number may be somewhat redundant, since it is difficult to think of any object other than a (real) number to which the second part of the statement applies.)

The *open interval* $(a, b)$ is defined to be the set of all numbers $x$ such that $a < x < b$, assuming that $a$ and $b$ are numbers such that $a < b$. Likewise the *closed interval* $[a, b]$ is the set of numbers $x$ such that $a \leqq x \leqq b$, where the symbol $\leqq$ means "is less than or is equal to." This enables us to make such concise statements as

$$t \text{ varies over } (0, 1)$$

in speaking of the variable $t$ in the previous paragraph.

It is often helpful to give a graphical representation of a number interval by emphasizing that portion of a scaled line whose points have as coordinates the numbers of the given interval. For instance, the *semiclosed* interval $0 < x \leqq 70$, or $(0, 70]$, is shown graphically in the upper part of Figure 1.3. Note how the solid dot at the point whose coordinate is 70 shows the inclusion of the point, while the open-centered dot at the left indicates that the point is omitted. The lower part of Figure 1.3 illustrates the closed interval $-3 \leqq x \leqq 3$, or $[-3, 3]$. Verify that this interval can also be indicated by the statement $|x| \leqq 3$.

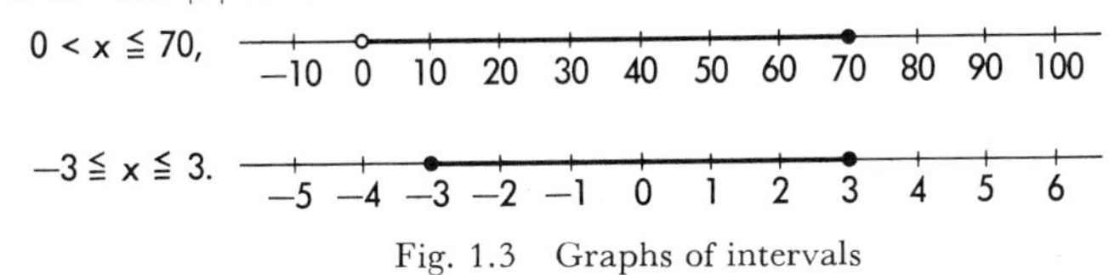


Fig. 1.3 Graphs of intervals

## *EXERCISES*

1. Assume that $P$ and $Q$ are distinct points. Indicate what each of the symbols $\overline{PQ}$, $\overrightarrow{PQ}$, $\overleftrightarrow{PQ}$, $PQ$, $\overrightarrow{QP}$, $QP$, $\overleftrightarrow{QP}$, $\overline{QP}$ represent. Which are sets and which are numbers?
2. **(a)** Which pairs of the symbols in Exercise 1 can properly be related by writing the symbol $=$ between them?
   **(b)** Which pairs of these symbols can properly be related by writing the symbol "$\subset$" between them?
3. Using Formula (1.1) and the scaled line shown below, find the indicated lengths of segments. Also, using the observation of the final paragraph of Section 1.4, determine the direction of the second point from the first in each pair. Check both the distance and direction by counting units on the scale.

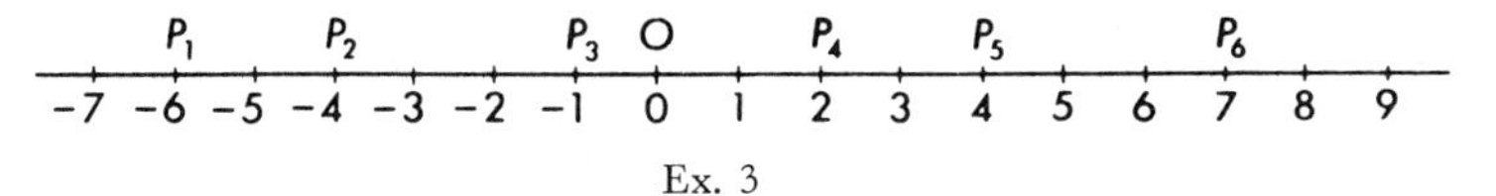


Ex. 3

(a) $P_5P_1$. (**Solution:** $x_1 - x_5 = (-6) - 4 = -10$, which is negative, so that $P_1$ is in the negative direction from $P_5$. Also $|x_1 - x_5| = 10$. Hence $P_5P_1 =$ length of $\overline{P_5P_1} = 10$. This checks because $P_1$ is 10 units to the left of $P_5$ on the figure.)
(b) $OP_1$ (c) $P_4P_6$ (d) $P_5P_2$ (e) $P_3P_6$ (f) $P_1O$

4. If $x_1$ and $x_2$ are coordinates of points $P_1$ and $P_2$ respectively, the formula

$$x = x_1 + \tfrac{3}{4}(x_2 - x_1)$$

gives the coordinate $x$ of the point $P$ that is three fourths of the way from $P_1$ to $P_2$. Try this formula to locate the point which is three fourths of the way from $P_1$ to $P_4$ in the figure of Exercise 3. Check by counting. Repeat to find the point three fourths of the way from $P_5$ to $P_2$.

5. In the formula of Exercise 4 replace the fraction $\tfrac{3}{4}$ by the fraction $\tfrac{1}{2}$ and simplify the result. Show (still using the figure of Exercise 3) that this formula correctly gives the coordinate of the midpoint of each of the following segments:

(a) $\overline{OP_4}$, (b) $\overline{P_1P_5}$, (c) $\overline{P_4P_6}$, (d) $\overline{P_4P_1}$.

6. Give graphical representations of each of the intervals given below.
(a) $-2 \leqq x \leqq 4$
(b) $0 \leqq x < 5$
(c) $x < 3$
(d) $|x| < 3$
(e) $[2, 4]$
(f) $(2, 4]$
(g) $|x| \geqq 3$, two intervals
(h) $x$ is greater than or equal to $-2$ but is less than 6

7. Express in inequality form the intervals determined by the following sketches. Indicate the same intervals using parentheses or braces.

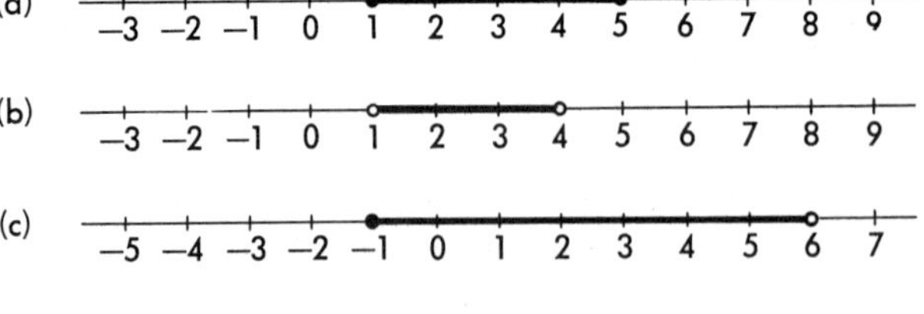


Ex. 7

8. The interval pictured in Exercise **7d** can be expressed as

$$\{x \mid x \text{ is a number and } |x| \leqq 4\}$$

Two of the other intervals indicated in Exercise 7 can similarly be expressed with the aid of absolute value symbols. Try to find which two and display the result.

---

## 1.6 A RECTANGULAR COORDINATE SYSTEM

In the same way that the position of a certain building in a city may be given by saying that it is "four blocks east and three blocks south of the City Hall," the position of a point in a plane can be described with reference to a pair of perpendicular* coordinate lines in the plane. Let these lines be so scaled that their point of intersection is the origin on each scale. Let the coordinate of a point on one line be the number $x$, and that of a point on the other line, $y$. These scaled lines are called the $x$ axis and the $y$ axis, respectively. The $x$ axis is usually placed in a horizontal position, as in Figure 1.4. The positive

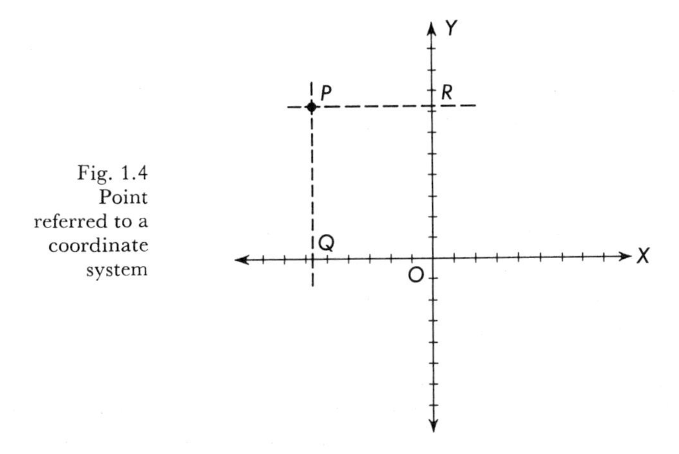


Fig. 1.4 Point referred to a coordinate system

directions are chosen as in Section 1.5. The arrowheads indicate the positive directions. Any point of the plane can be identified by coordinates on these axes as follows. A line through $P$ parallel to the $y$ axis will intersect the $x$ axis at a point $Q$ with $x$ coordinate $x$. Similarly, a line through $P$ parallel to the $x$ axis will intersect the $y$ axis at a point $R$ with $y$ coordinate $y$. In this way, to every point $P$ of the plane there corresponds a unique pair of coordinates $x$ and $y$ that are known as the *coordinates* of $P$. No other point has the same pair of coordinates. Thus the symbol $(x, y)$ is used to identify the point $P$.

---

* A pair of nonperpendicular intersecting lines may also be made the basis of a coordinate system, but a rectangular system is more useful for the purposes of this book. Any such system is called a Cartesian coordinate system, after the French mathematician, R. Descartes (1596–1650).

When desired, the letter identification and the coordinate-pair identification can be united in the form $P(x, y)$. Note that the $x$ coordinate of $P$ is the directed distance $OQ = PR$ and that the $y$ coordinate of $P$ is $OR = QP$. The $x$ coordinate and the $y$ coordinate of a point are known respectively as the *abscissa* and the *ordinate* of the point.

The plane is divided into four *quadrants* by the coordinate axes. These are numbered with Roman numerals in counterclockwise fashion with the number-one quadrant in the upper right-hand position. (See Figure 1.5.)

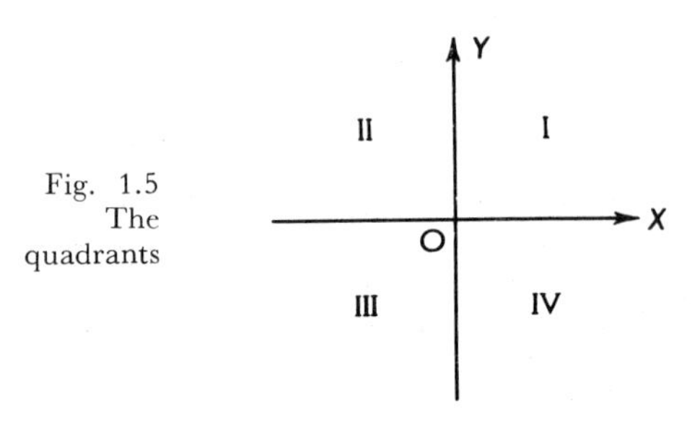


Fig. 1.5 The quadrants

## EXERCISES

**1.** Construct a rectangular coordinate system and locate the following points: $A(2, 3)$, $B(-4, 2)$, $C(0, 4)$, $D(-2, 0)$, $E(-3, -5)$, $F(-1, \sqrt{3})$, $G(\sqrt{2}, -1)$, $O(0, 0)$.

**2.** If $(x, y)$ represents a variable point, state
(**a**) the sign of $x$ in each quadrant,
(**b**) the sign of $y$ in each quadrant,
(**c**) the sign of $y/x$ in each quadrant.

**3.** Shade the region that consists of the points $(x, y)$ satisfying the inequality $x \geqq 2$. (*Note:* Since $y$ is not mentioned, it may take on all possible values.)

## 1.7 OTHER PLANE FIGURES

We have discussed lines, rays, and line segments as figures (sets of points) in a plane. Other familiar plane figures of particular interest in trigonometry are *angles* (to be discussed fully in Chapter 2), *triangles*, and *circles*. We assume familiarity with these figures and their properties pausing for a brief review.

In geometry, an *angle* is the union of two rays having a common end point, called the *vertex* of the angle. Angles may be measured in degrees or in other units of angle measure that will be discussed later.

If $A$, $B$, $C$ are three points not all on one line, then the triangle $ABC$, or $\Delta ABC$, is the set of points formed by the union of segments $\overline{AB}$, $\overline{BC}$, and

$\overline{CA}$. To say that $\Delta ABC$ is *congruent* to $\Delta XYZ$ means that $AB = XY$, $AC = XZ$, $BC = YZ$, and that the measures of angles $A$, $B$, and $C$ are equal to the respective measures of angles $X$, $Y$, and $Z$. The reader will recall the *s.a.s.* = *s.a.s.* condition and others that assure congruence of triangles with less known of them than is called for by the definition of congruence.

In trigonometry there is particular interest in *similar* triangles. Two triangles are similar if their corresponding sides have proportionate lengths (meaning that, the ratios of the lengths of corresponding sides are equal) and the corresponding angles are of equal measure. This definition also applies to polygons. However, similarity of triangles is assured if it is known either that the corresponding sides are proportional or that the corresponding angles have equal measures. More than this, since the sum of the measures of the three angles of any triangle is 180 (degrees) similarity follows if two corresponding angles are known to be equal. (Figure 1. 6.)

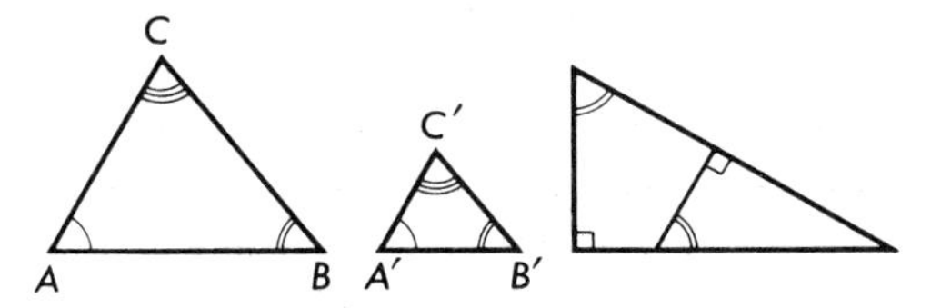


Fig. 1.6 Similar triangles

The *circle* with center 0 and (length of) radius $r$ is the set of all points* whose distance from 0 is $r$. Or, if $C(O, r)$ represents the phrase "the circle with center 0 and radius $r$," then

$$C(0, r) = \{P \mid P \text{ is a point and } OP = r\}$$

Figure 1.7 illustrates various terms used in connection with the circle.

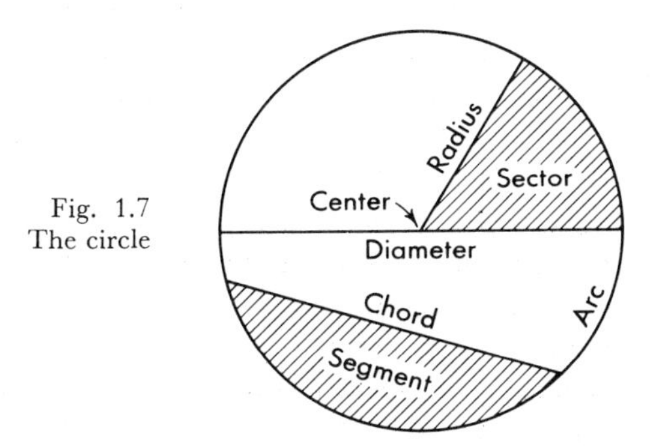


Fig. 1.7
The circle

A *secant to a circle* is a line whose intersection with the circle consists of two points, as line $PQ$ in Figure 1.8. A *tangent to a circle* is a line whose inter-

* Remember, we are considering only points of a plane.

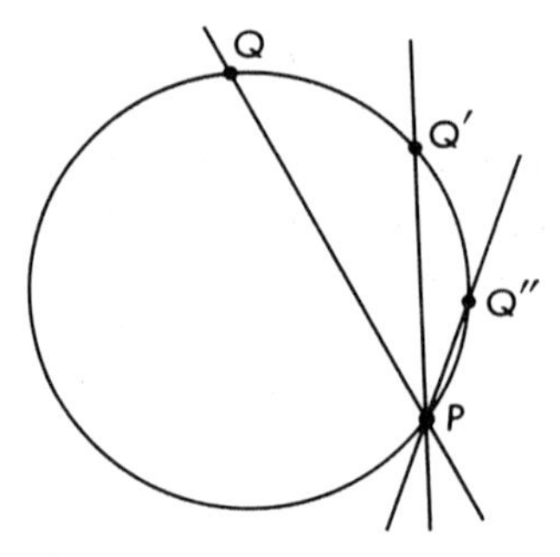


Fig. 1.8 Secants to a circle

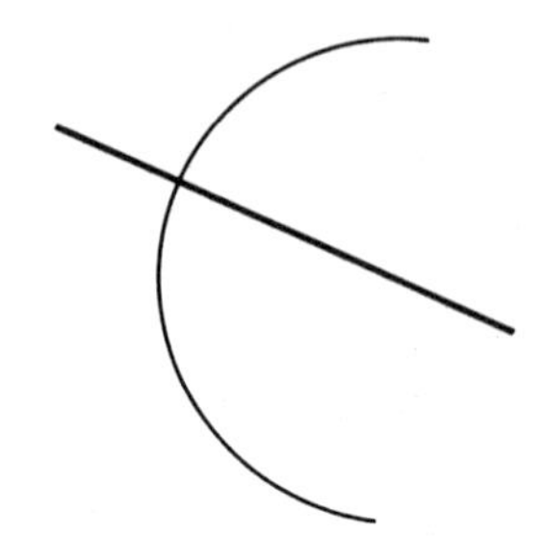

Fig. 1.9 Line with one point in common with a curve

section with the circle is a single point. Note that this definition does not always say what we want it to if we are talking about a tangent to a curve that is not a circle. For example, this definition fails for as simple curve as a half circle (see Figure 1.9). As a suggestion for how a more satisfactory definition of a tangent to a curve may be developed, consider what happens to the secant line $QP$ (Figure 1.8) if $Q$ is moved around the circle toward $P$. We observe that the secant line rotates about point $P$. The nearer $Q$ moves toward $P$ the more does line $QP$ look like what we expect the tangent line at $P$ to be. We rather casually say that the tangent line at $P$ is the limiting position of the line $PQ$ as $Q$ approaches $P$, assuming there *is* such a limiting position. This concept of the tangent line, more carefully developed in the subject of calculus, applies to all types of curves.

## *EXERCISES*

1. A building with vertical sides stands on level ground. If it casts a shadow 72 ft long at the same time that a post 6 ft tall casts a shadow 4 ft long, how tall is the building?
2. The distance from home plate to second base on a standard baseball diamond 90 ft square is approximately 127 ft. How far is it from home plate to second base on a junior-size diamond 60 ft square?
3. Sketch an angle whose measure is less than 90°. Let 0 be the vertex of the angle and let $A$ and $A'$ be points on one side of the angle. From $A$ and $A'$ drop perpendiculars meeting the other side of the angle at $B$ and $B'$, respectively. Show that $\triangle OBA$ is similar to $\triangle OB'A'$ and then complete each of the following proportions: **(a)** $\dfrac{BA}{B'A'} =$ ; **(b)** $\dfrac{OB}{AB} =$ **(c)** $\dfrac{AB}{OA} =$

**4.** A street light is 27 ft above level ground. How long is the shadow of a 6-ft man who stands 18 ft from the point directly under the light?
**5.** Repeat Exercise 4, except that the man stands $x$ ft from the point under the light, expressing the length of the shadow in terms of $x$.
**6.** A ladder 20 ft long leans against a vertical wall, the top of the ladder reaching 16 ft up the wall. One rung of the ladder is 15 ft from the lower end of the ladder. What is the vertical height of this rung?
**7.** The legs (perpendicular sides) of a certain right triangle are 11 in. and 25 in. in length, respectively. From a point on the hypotenuse, perpendiculars are let fall to the legs of the triangle, forming a rectangle whose length is 15 in. Find the width of the rectangle.
**8.** Show by a sketch what is wrong with the following attempt at a definition of a line tangent to a curve: "A line is tangent to a curve at a point $P$ if it touches the curve at $P$ but does not cross the curve there."

---

## 1.8 THE PYTHAGOREAN THEOREM

One of the most famous and widely applied laws of plane geometry is the right-triangle theorem attributed to the Greek mathematician and philosopher Pythagoras (582–507 B.C.). It is the familiar truth that *the square upon the hypotenuse of any right triangle is equal to the sum of the squares upon the two legs*, or, if $c$, $a$, and $b$ denote the lengths of the hypotenuse and two legs, respectively, then

$$c^2 = a^2 + b^2 \tag{1.2}$$

The converse is also true: if the sides of a triangle are such that $c^2 = a^2 + b^2$, then the triangle is a right triangle with right angle opposite the side of length $c$.

If two of the three numbers $a$, $b$, and $c$ are known, Equation (1.2) may be used to evaluate the third one. We thus have $c = \sqrt{a^2 + b^2}$; $a = \sqrt{c^2 - b^2}$; and $b = \sqrt{c^2 - a^2}$. We here remind the reader of the convention (rule by agreement) that the symbol $\sqrt{x}$ is to be read "the non-negative square root of $x$," assuming $x \geqq 0$. Thus

$$\sqrt{9} = 3, \textit{ not } \pm 3$$

If we wish to indicate the negative square root of a number $x$, we write $-\sqrt{x}$. We further note that

$$\sqrt{x^2} = |x|$$

so that the $\sqrt{\ }$ sign is an effective symbol of absolute value when used this way.

 **EXAMPLE 1**

Find the radius of a regular octagon each side of which is 10 in.

**Solution:** By extending two perpendicular sides of the octagon an isosceles right triangle will be formed (see Figure 1.10). Let $x$ equal the length of each

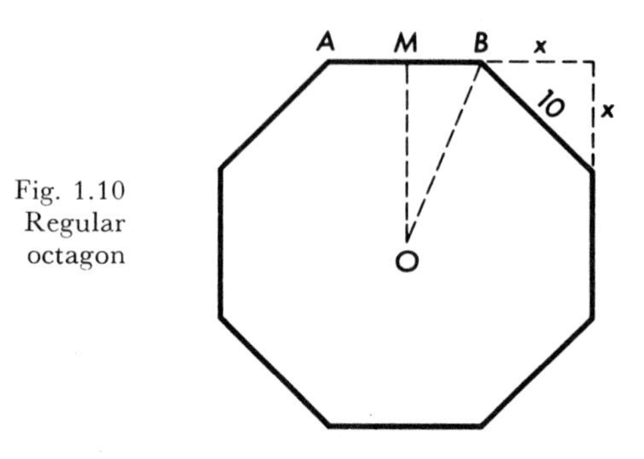


Fig. 1.10 Regular octagon

leg, so that

$$x^2 + x^2 = 10^2$$

from which

$$x^2 = 50$$

and

$$x = 5\sqrt{2} \text{ units}$$

Now, if $M$ denotes the midpoint of $AB$,

$$OM = x + 5 = 5(1 + \sqrt{2})$$

Since

$$\begin{aligned} OB^2 &= OM^2 + MB^2 \\ &= 25(1 + \sqrt{2})^2 + 25 \\ &= 25(1 + 2\sqrt{2} + 2 + 1) \\ &= 50(2 + \sqrt{2}) \end{aligned}$$

it follows that

$$OB = 5\sqrt{4 + 2\sqrt{2}} \text{ units}$$

 **EXAMPLE 2**

A ship sailed 15 miles eastward and then 12 miles northward. How far was it then from the starting point?*

* Unless otherwise stated, triangles on the earth's surface will be treated in this book as plane triangles. This assumption will give fairly accurate results for triangles of small size (sides of a few miles). Spherical trigonometry is needed to deal with actual spherical triangles.

**Solution:** The unknown distance is the length of the hypotenuse of a right triangle whose legs are 12 and 15. Hence, by the Pythagorean law,

$$d^2 = 12^2 + 15^2$$
$$= 369$$
$$d = \sqrt{369} = 19.2 \text{ miles, approximately}$$

**EXAMPLE 3**

Given that $x$ is a number such that $x \geqq 4$, simplify each of the following:

(a) $\sqrt{x^2}$; (b) $\sqrt{(x-4)^2}$; (c) $\sqrt{(4-x)^2}$.

**Solution:** Recall that for any number $a$, $|a| = a$ if $a \geqq 0$ and $|a| = -a$ if $a < 0$. Then

**A** $\sqrt{x^2} = |x| = x$, since $x > 0$.

**B** $\sqrt{(x-4)^2} = |x-4| = x-4$, since $x - 4 \geqq 0$.

**C** $\sqrt{(4-x)^2} = |4-x| = -(4-x)$, since $4 - x \leqq 0$. Thus the answer to (c) is 0 if $x = 4$ and is $x - 4$ if $x > 4$.

## *EXERCISES*

1. What is the value of $\sqrt{(x-3)^2}$ if $x = 5$? If $x = 1$?
2. Under what circumstances is $\sqrt{(a-x)^2} = a - x$?
3. On a scaled line show graphically the set of numbers $x$ such that $\sqrt{x} < 4$.
4. A chord of length 8 in. is drawn in a circle of radius 5 in. Find the distance from the center of the circle to the chord.
5. Find the base of an isosceles triangle if its perimeter is 50 ft and its altitude is 15 ft.
6. Find the side of an equilateral triangle if its altitude is 10 ft.
7. The Pythagorean Theorem and its converse assure us that a triangle is a right triangle if, and only if, the sum of the squares of the lengths of two of its sides is equal to the square of the length of the third side. Determine which of the following number triples can be lengths of the respective sides of a right triangle. Assume that the literal expressions represent positive numbers: (**a**) 3, 4, 5; (**b**) 7, 24, 25; (**c**) 8, 19, 23; (**d**) 2, $\sqrt{21}$, 5; (**e**) $5k$, $12k$, $13k$; (**f**) $x^2 - y^2$, $2xy$, $x^2 + y^2$.
8. Show that the Pythagorean Theorem may be proved by use of proportionate sides of similar triangles by filling in the details of the following:

**HYPOTHESIS**

$\Delta ABC$ is a right triangle with right angle at $C$. (See figure.)

**CONCLUSION**

$AC^2 + CB^2 = AB^2$

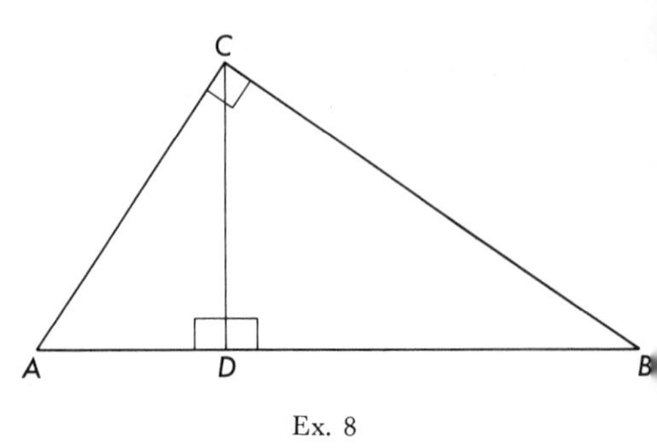


**PROOF**

Ex. 8

Let $\overline{CD}$ be the perpendicular from $C$ to $\overline{AB}$. $\Delta ADC$ is similar to $\Delta ACB$. (Why?)

$$\therefore \frac{AD}{AC} = \frac{AC}{AB}, \text{ so that } AD = \frac{AC^2}{AB} \text{ (Why?)}$$

$\Delta DBC$ is similar to $\Delta CBA$. (Why?)

$$\therefore \frac{DB}{CB} = \frac{CB}{AB}, \text{ so that } DB = \frac{CB^2}{AB} \text{ (Why?)}$$

Adding the expressions for $AD$ and $DB$, gives,

$$AD + DB = \frac{AC^2 + CB^2}{AB}$$

(Completion of the proof is left to the student.)

**Note:** This proof tacitly assumes that $D$ is between $A$ and $B$. A proof that $D$ *is* between $A$ and $B$ can be found in modern geometry texts.

**9.** The distance that an observer in an elevated position can see in level country or at sea may be defined as the length of a tangent drawn from the observer to the earth's surface. Show that this distance $d$, *in miles*, for an observer elevated to a height $h$, *in feet*, is given approximately for

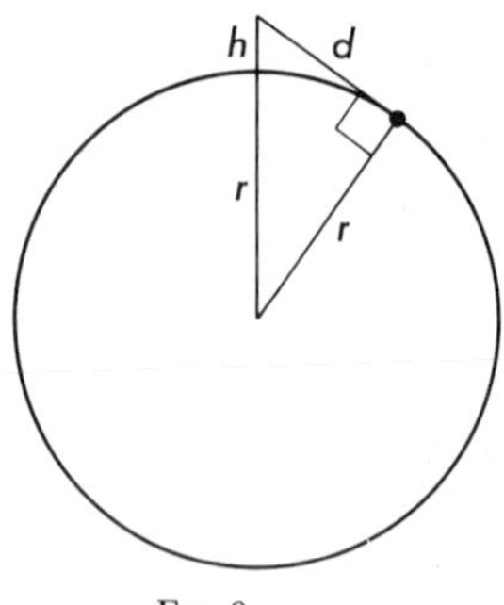


Ex. 9

small values of $h$ by the formula

$$d \approx \sqrt{\frac{3h}{2}},$$

where the symbol "$\approx$" is read "is approximately equal to."
(*Suggestion:* The radius of the earth is close to 3960 miles. $h$ feet $= h/5280$ miles. Apply the Pythagorean rule to the right triangle suggested by the figure and solve the resulting equation for $d$ in terms of $h$. In the result, delete the term $h^2/5280$, since it will be seen to be negligible for relatively small values of $h$.)

**10.** Apply the formula of Exercise 9 to approximate the greatest distance at which an observer 200 ft above the surface of the ocean can see an object at water level.

**11.** What is the greatest distance at which the observer of Exercise 10 can see an object which is 100 ft above water level?

## 1.9 THE DISTANCE BETWEEN TWO POINTS

The distance between two points $P_1(x_1, y_1)$ and $P_2(x_2, y_2)$ can be found by using the Pythagorean law. By drawing lines through $P_1$ and $P_2$ parallel to the coordinate axes, a right triangle is formed with the vertex of the right

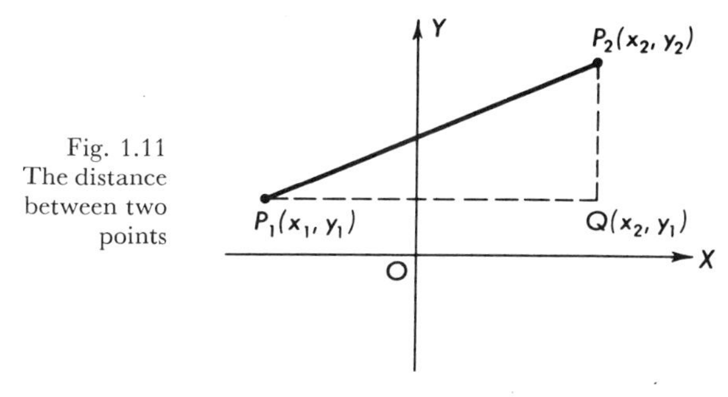


Fig. 1.11 The distance between two points

angle at $Q(x_2, y_1)$, as in Figure 1.11. If $d$ represents the distance $P_1P_2$, then

$$d^2 = P_1Q^2 + QP_2^2$$

But, from formula (**1.1**)

$$P_1Q^2 = (x_2 - x_1)^2 \quad \text{and} \quad QP_2^2 = (y_2 - y_1)^2$$

Hence

$$d^2 = (x_2 - x_1)^2 + (y_2 - y_1)^2$$

or

$$d = \sqrt{(x_2 - x_1)^2 + (y_2 - y_1)^2} \qquad \textbf{(1.3)}$$

Note that this formula is valid regardless of the relative positions of $P_2$ and $P_1$ in the plane, even when they are on the same vertical or horizontal line. (The student should check this.)

**EXAMPLE**

Find the distance between the points (3, 4) and (−5, 1).

**Solution:** Let $P_1$ and $P_2$ be (3, 4) and (−5, 1), respectively. Then,

$$\begin{aligned} d &= \sqrt{(-5-3)^2 + (1-4)^2} \\ &= \sqrt{(-8)^2 + (-3)^2} \\ &= \sqrt{73} \text{ units} \end{aligned}$$

## *EXERCISES*

1. Repeat the solution of the above example using $P_1 = (-5, 1)$ and $P_2 = (3, 4)$. What is your conclusion? Make your conclusion general by an examination of formula (1.3) with the order of points reversed.
2. Find the distance between each of the following pairs of points. As a partial check, plot the pair of points on a rectangular coordinate system and judge whether or not your result seems reasonable.

   (**a**) (−2, 1), (3, −4)
   (**b**) (0, −2), (−2, 0)
   (**c**) (3, −1), (−1, −4)
   (**d**) (−2, 1), (3, 1)

3. Show that the points (2, 3), (1, 6), and (−2, 7) are the vertices of an isosceles triangle.
4. Show that the points (0, −3), (0, 3), and $(-3\sqrt{3}, 0)$ are vertices of an equilateral triangle.
5. Show that the points (−3, −1), (5, −2), and (3, 2) are the vertices of a right triangle.
6. Draw a circle with center (5, 3) that is tangent to the $y$ axis. Does the circle pass through the point (1, 0)? (2, 7)? (5, 9)? (8, −1)? Do not rely on your drawing alone for your answers.
7. Describe the set of all points $(x, y)$ such that $\sqrt{(x-2)^2 + (y-3)^2} = 4$. Is this the same as the set $\{(x, y) \mid (x-2)^2 + (y-3)^2 = 16\}$?
8. Shade the region consisting of the points that satisfy each of the following conditions: (**a**) $x^2 + y^2 \leqq 16$; (**b**) $\sqrt{(x-1)^2 + (y+2)^2} \geqq 2$; (**c**) $x^2 + (y-3)^2 \leqq 1$; (**d**) $1 \leqq (x-2)^2 + (y-1)^2 \leqq 9$.

## MISCELLANEOUS EXERCISES/CHAPTER 1

1. If $p$ and $q$ are positive integers such that $p > q$, show that the numbers $a$, $b$, and $c$ given by the formulas $a = p^2 - q^2$, $b = 2pq$, and $c = p^2 + q^2$ are integers such that the triangle with sides of lengths $a$, $b$, and $c$ is a right triangle. (Such a set of three integers is called a Pythagorean number triple.)
2. Find three distinct Pythagorean number triples that you had not noticed before by selecting different values of $p$ and $q$ and using the formulas of Exercise 1.
3. Show that if $p$ and $q$ differ by 1, it always follows that two of the three resulting Pythagorean numbers also differ by 1.
4. In each of the following, use the distance formula to express the distance between the given pair of points. Simplify the result, if possible.

   (**a**) $(4, -5)$, $(-2, 3)$
   (**b**) $(3, 6)$, $(-2\sqrt{2}, \sqrt{2})$
   (**c**) $(7k, k)$, $(4k, -3k)$
   (**d**) $(15, 0)$, $(0, 8)$

5. Let $A$ be the set $\{x \mid x \text{ is a number and } 0 < x \leqq 4\}$, and let $B$ be the set $\{x \mid x \text{ is a number and } 0 \leqq x < 5\}$.

   (**a**) Show these sets graphically on a number line.
   (**b**) Show graphically $A \cap B$ and $A \cup B$.
   (**c**) Express $A \cap B$ and $A \cup B$ in the form that $A$ and $B$ are expressed in at the beginning of this exercise.

6. (**a**) On a coordinate system plot two points and name them $P_1(x_1, y_1)$ and $P_2(x_2, y_2)$. Show that the point $P(x, y)$, where

$$x = x_1 + \tfrac{3}{4}(x_2 - x_1) \quad \text{and} \quad y = y_1 + \tfrac{3}{4}(y_2 - y_1)$$

   is located on $\overline{P_1P_2}$ three fourths of the way from $P_1$ to $P_2$. (*Suggestion:* consider the similar triangles that can be constructed as shown in the figure.)

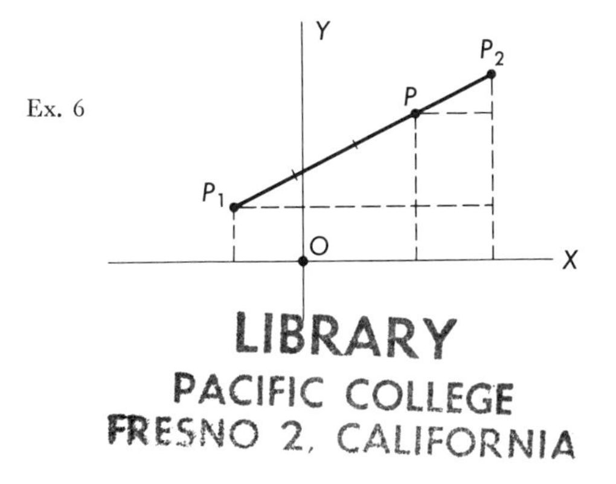

Ex. 6

(**b**) Find the point three fourths of the way from $(-2, 1)$ to $(6, 5)$.

**7.** Using the fact that $a > b$ if and only if $a - b > 0$ (that is, if $a - b$ is positive), show that

(**a**) $x + 1 > x$, for every number $x$
(**b**) $-2 > -7$
(**c**) $x^2 + y^2 \geqq 2xy$, for every number $x$ and every number $y$

**8.** The points $A(-2, 5)$, $B(1, 6)$, and $C(-1, 2)$ are the vertices of a triangle. Which of the following statements are true and which are false? Justify your answer. (**a**) $\triangle ABC$ is isosceles; (**b**) $\triangle ABC$ is a right triangle; (**c**) $\triangle ABC$ is equilateral.

**9.** Three points are known to lie on a straight line if $AB + BC = AC$, where $A$, $B$, and $C$ are names of the points in some order. Show that the points $(-1, 2)$, $(5, 6)$, and $(-4, 0)$ lie on a line. Check with a sketch.

**10.** $P_1(x_1, y_1)$ and $P_2(x_2, y_2)$ are points on a ray whose end point is the origin of the coordinate system. Suppose that $P_1$ and $P_2$ are distinct from each other and the origin. Let $r_1$ and $r_2$ be the respective distances $OP_1$ and $OP_2$. Verify the following proportions:

$$\frac{x_1}{r_1} = \frac{x_2}{r_2} \qquad \frac{y_1}{r_1} = \frac{y_2}{r_2} \qquad \text{and} \qquad \frac{y_1}{x_1} = \frac{y_2}{x_2}$$

(*Hint:* Consider the similarity of the right triangles formed by drawing perpendiculars to the $x$ axis from $P_1$ and $P_2$, as in the illustration.)

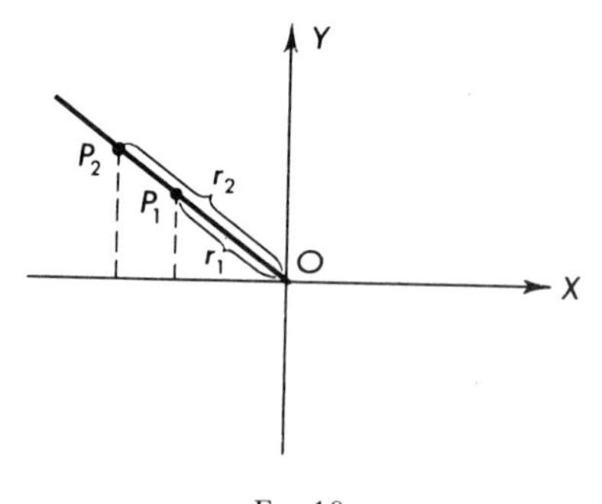


Ex. 10

**11.** *A mettle tester*. The perimeters of a certain regular hexagon and a certain regular octagon are equal. Find the ratio of the area of the octagon to the area of the hexagon.

# CHAPTER 2. THE TRIGONOMETRIC FUNCTIONS

## 2.1 DIRECTED ANGLES

In geometry an angle is the figure formed by the union of two rays having the same end point and not lying on the same line. In trigonometry we shall regard an angle as being generated by rotating one of the rays about its end point into the position of the other. Thus we define an angle to be *the geometric angle together with the rotation used to bring one side (the initial side) into the position of the other side (the terminal side)*. The end point about which the ray rotates is the *vertex*. The angle is regarded as *positive* if the rotation is *counterclockwise* (opposite to the direction of motion of the hands of a clock), *negative* if the rotation is *clockwise*. A curved arrow (Figure 2.1) serves to indi-

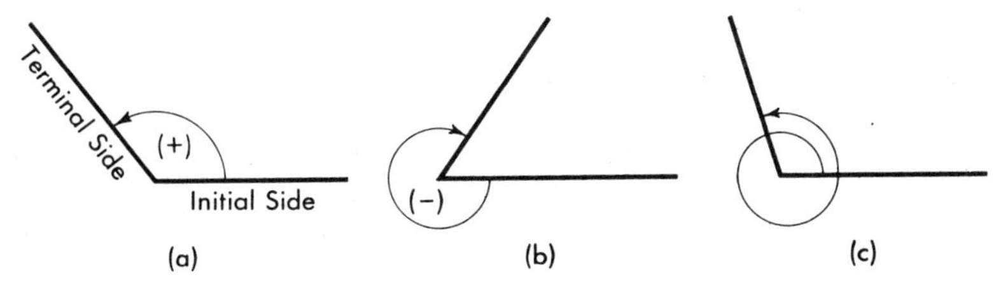


Fig. 2.1 (a) A positive angle (b) A negative angle (c) An angle of more than revolution

cate the direction and amount of rotation. The angle thus defined is not confined to any range of magnitude. A spoke of a rotating flywheel may generate an angle of thousands of revolutions in a few minutes of time.

To each angle in the set of all angles is assigned a number called its *measure*. If the unit of angle measure is a *revolution*, the number assigned as the measure of an angle is the number of revolutions of rotation from the initial side to the terminal side. If the unit of angle measure is the *degree*, which is 1/360 of a revolution, the number assigned as the measure of an angle is the number of degrees involved in the rotation. Another unit of angle measure, the *radian*, is of importance in calculus. It will be discussed in Chapter 3. The idea we are stressing here is that, given a unit of measure, to each angle is assigned a number that is its measure. This number is positive when the angle is positive and negative when the angle is negative, so that the measure of an angle serves to indicate the direction as well as the magnitude of the rotation involved.

In this chapter we shall employ degree measure of angles. The familiar *minute* is 1/60 of a degree. The *second* is 1/60 of a minute. It would be difficult to draw an angle of one second on a sheet of paper. At a distance of one mile from the vertex, the sides of an angle of one second are separated by a distance approximately equal to the thickness of a lead pencil. Delicate instruments are required to measure angles accurately to the nearest minute. The customary symbols, $^\circ$ for *degree*, $'$ for *minute*, and $''$ for *second*, will be used in this book.

Angles that have equal measures we shall call *equal* angles.* A single Greek letter will frequently be used to represent an angle. The symbol "$m(\theta)$" is then read "the measure of angle $\theta$." The symbol "$60^\circ$" is read "angle of 60 deg." Note that $m(60^\circ) = 60$, it being obvious that degree measure is being used. In keeping with our notation, we may write such statements as the following:

"If $\theta$ is an angle such that $m(\theta) = 75$ (degree measure), then

$$\theta = 75^\circ\text{''}$$

If angle $\theta$ is such that $-90 < m(\theta) < 90$, $\theta$ is called an *acute* angle.

## 2.2 STANDARD POSITION OF AN ANGLE

An angle is said to be in standard position with reference to a rectangular coordinate system if the origin of the coordinate system is at the vertex of the angle and if the positive half of the $x$ axis forms the initial side of the angle. In Figure 2.2, angles of $+150^\circ$ and $-70^\circ$ are shown in standard position. Note again that the positive angle involves a counterclockwise rotation, while the negative angle involves a clockwise one.

A partial description of an angle in standard position is a statement of the quadrant (see Figure 1.8) in which the terminal side lies. Thus we say that an angle of $150^\circ$ *terminates in the second quadrant*, or that an angle of $150^\circ$ is a *second-quadrant angle*. Similarly, an angle of $-70^\circ$ is a fourth-quadrant angle.

---

* A current trend in geometry is to use the word *congruent* for angles of equal measure. We somewhat reluctantly retain the word *equal*, because this use is so imbedded in the literature of applications of trigonometry in science, and so forth, that familiarity with its use seems desirable.

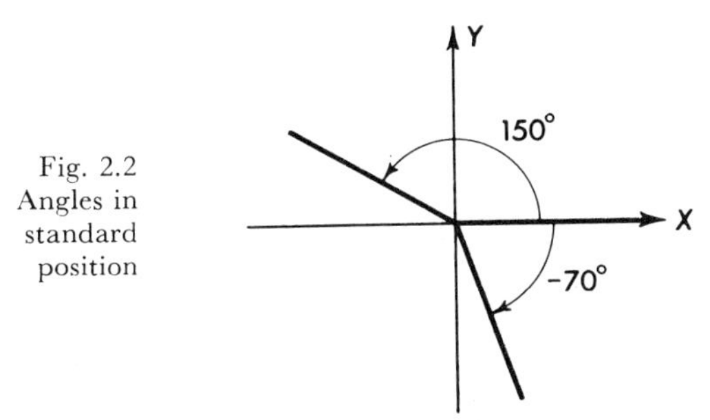


Fig. 2.2 Angles in standard position

If two or more angles in standard position have the same terminal side, they are said to be *co-terminal*. For example, angles of 225°, −135°, and 585° are co-terminal.

**EXAMPLE**

Find the smallest positive angle that is co-terminal with (the angle of) 1356°.

**Solution:** By division by 360, we find that $1356 = (3)(360) + 276$. Hence the smallest positive angle co-terminal with 1356° is 276°. (See Figure 2.3.)

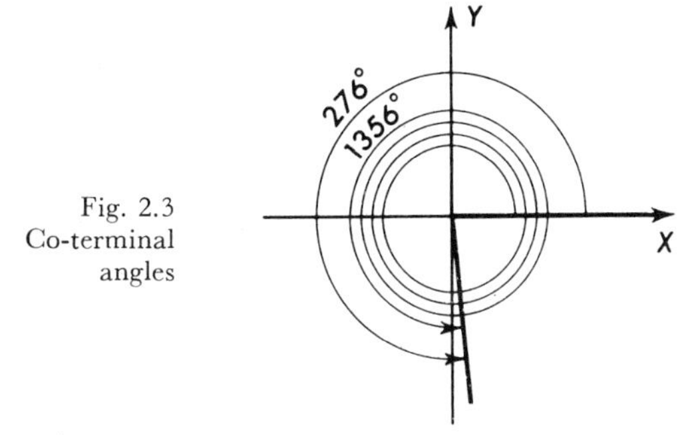


Fig. 2.3 Co-terminal angles

## *EXERCISES*

**1.** Construct each of the following angles in standard position, give the quadrant in which it terminates, and name one other angle co-terminal with it.

(**a**) 35°
(**b**) −150°
(**c**) 410°
(**d**) 750°
(**e**) −85°
(**f**) 270°
(**g**) −180°

**2.** Give a negative angle that is co-terminal with each of the following.

(**a**) $90°$
(**b**) $180°$
(**c**) $0°$
(**d**) $360°$

**3.** Construct an angle $\theta$ in standard position with its terminal side passing through the point $(-3, 3)$. Give two values of $\theta$ subject to the condition that $-360 < m(\theta) < 360$.

**4.** Find the smallest positive angle that is co-terminal with $-480°$. Find the angle with measure of smallest *absolute value* that is co-terminal with $-480°$.

**5.** Find an angle $\theta$ such that $720 < m(\theta) < 1080$ that is co-terminal with $250°$.

**6.** Find the angle generated by a clock hand in 3 hr and 40 min if the hand is (**a**) the minute hand, (**b**) the hour hand. (Be sure to give the sign of these angles.)

**7.** Two angles are said to be *complementary* if the sum of their measures is 90. Each is called the *complement* of the other. Find the complement of each of the following.

(**a**) $62°$
(**b**) $-10°$
(**c**) $430°$

**8.** If the sum of the measures of two angles is 180, they are said to be *supplementary* and each is the *supplement* of the other. Find the supplement of

(**a**) $50°$
(**b**) $90°$
(**c**) $-90°$

---

## 2.3 FUNCTIONS

We are frequently concerned with relating the elements of one set with elements of another set. In Chapter 1, we assigned to each point of a line a number, thus relating the set of points on a line to the set of real numbers. Again to each angle we assign a number as its measure and thus relate the set of all angles to the set of real numbers. These are examples of particular types of relations between sets known as *functions*. Let us now give a definition.

*If with each element of a set $A$, there is associated in some way exactly one element of another set $B$, then this association constitutes a function from $A$ to $B$.*

There are thus three essentials in order to have a function:

(1) A set $A$, called the *domain* of the function.
(2) A set $B$, called the *range* of the function.
(3) A correspondence that associates with each element $x$ of $A$ exactly one element $y$ of $B$.

A frequently used symbolism for a function from $A$ to $B$ is

$$f: A \rightarrow B$$

If $y$ is the element of $B$ that $f$ associates with a given element $x$ of $A$, we write

$$y = f(x), \text{ read "} y \text{ is } f \text{ of } x.\text{"}$$

Note that the sets $A$ and $B$ of a function are not restricted as to the number of elements they contain. For instance, let $A = \{a, b, c\}$ and $B = \{1, 2\}$. Suppose the elements of $A$ are paired with those of $B$ as follows:

$$a \rightarrow 1, \quad b \rightarrow 2, \quad c \rightarrow 1$$

We see that this a function $f: A \rightarrow B$, since there is a domain $A$, a range $B$, and a correspondence associating each element of $A$ with exactly one element of $B$. We observe that $f(a) = 1$, $f(b) = 2$, and $f(c) = 1$. Note also that while two elements of $A$ are associated with the same element of $B$, we still have the required condition that *each* element of $A$ is associated with *exactly one* element of $B$. Of course there is nothing sacred about the letters $A$, $B$, and $f$ in discussing functions. Other letters are often more suggestive. For example: $l: P \rightarrow R$, where $P$ is the set of all positive numbers and $R$ is the set of real numbers and the correspondence is such that if $x \in P$, then $l(x) = \log_{10} x$ is the corresponding element of $T$, is a function with domain $P$ and range $R$.

If both the domain and range of a function are sets of real numbers, the function assigns to each number $x$ belonging to the domain a number $y$ of the range (though not all of the elements of the range need be assigned). These ordered pairs of numbers of the form $(x, y)$ may be plotted on a rectangular coordinate system. The resulting set of points is called the *graph of the* function. Indeed, if the ordered pairs are regarded as *being the points*, the set of points *is* the function. (Its domain is the set of $x$ coordinates, its range is the set of $y$ coordinates, and the rule of correspondence determines $y = f(x)$ for each choice of $x$ in the domain.) We return to the matter of graphs of functions in Chapter 4.

## *EXERCISES*

**1.** Given the function $f: A \rightarrow B$, where $A$ and $B$ are the real numbers and $f(x) = x^2 - x$, find

(**a**) $f(2)$ (**b**) $f(0)$ (**c**) $f(k)$ (**d**) $f(-3)$

**2.** Which of the following sets of ordered pairs constitute a function $f$? Assuming all ordered pairs of a function are given, give the domain and range.

(**a**) $\{(1, 2), (2, 4), (3, 6), (4, 7)\}$
(**b**) $\{(1, 1), (2, 1), (3, 1)\}$
(**c**) $\{(1, 1), (1, 2), (1, 3)\}$
(**d**) $\{(-1, 0), (3, -1), (0, 0)\}$

**3.** If the elements of the domain and the range of a function are to be *integers* and $f(x) = \sqrt{25 - x^2}$, write all of the ordered pairs of integers that the function produces.

**4.** Is the following a function?

$$d: A \rightarrow B$$

where $A$ is the set of all *pairs of points* in a plane, $B$ is the set of non-negative numbers, and $s$ assigns to each pair of points in $A$ the number in $B$ that is the distance between the points of the pair? Justify your answer.

**5.** Is the following a function?

$$s: M \rightarrow S$$

where $M$ is the set of mothers in Chicago who have sons living in Chicago, $S$ is the set of sons in Chicago, and if $x \in M$, then $s(x) =$ son of $x$. Justify your answer.

## 2.4 THE TRIGONOMETRIC FUNCTIONS

Let $\angle XOT$ be any angle $\theta$ in standard position, and let $P(x, y)$ be an arbitrarily chosen point (different from the origin) on the terminal side of $\theta$. (See Figure 2.4.) Let $r$ represent the distance $OP = \sqrt{x^2 + y^2}$. With each such angle let us associate the number $y/r$. By a consideration of similar triangles (See Exercise 10 of Miscellaneous Exercises, Chapter 1), it is seen that the number $y/r$ does not depend upon the choice of the point $P$ on the terminal side of $\theta$. Hence to each choice of $\theta$ there is associated exactly one number, the ratio $y/r$. Thus we have a function

$$s: A \rightarrow R$$

in which $A$ is the set of all angles in standard position, $R$ is a set of real numbers, and if $\theta \in A$, then $s(\theta) = y/r$. The domain of the function is the set of all angles,* and the range is the set of real numbers.

* Since the *measure of an angle* completely identifies the angle in standard position, the domain could be the set of measures (the real numbers) of all angles. We shall discuss this in Chapter 3.

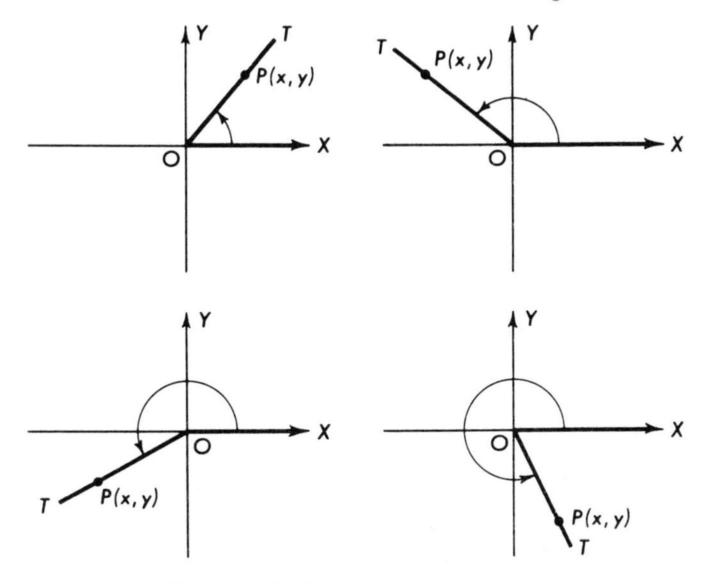


Fig. 2.4 Angles in the various quadrants

The function we have described in the preceding paragraph is called the *sine* function. Let us repeat the description, replacing the letter $s$ by the word sine or its abbreviation sin. Thus

$$\sin: A \rightarrow R$$

is a function from the set of angles $A$ to the real numbers $R$ such that to each $\theta$ belonging to $A$ is assigned the number $\sin \theta$ given by the equation

$$\sin \theta = \frac{y}{r}$$

where $y$ and $r$ are the abscissa and distance to the origin respectively of an arbitrary point on the terminal side of $\theta$ (different from the origin), assuming that $\theta$ is in standard position with respect to a rectangular coordinate system.

In addition to the sine function, there are five other named functions associating angles with real numbers. These are the *cosine*, *tangent*, *cotangent*, *secant*, and *cosecant* functions that are defined below. These with the sine function are known as the *trigonometric functions*. They form the basis of the subject of trigonometry and should be committed to memory. Using $A$ and $R$ as the sets associated with the sine function, the definitions of the five functions are (note abbreviations):

cosine: $A \rightarrow R$, with $\cos \theta = \dfrac{x}{r}$

tangent: $A' \rightarrow R$, with $\tan \theta = \dfrac{y}{x}$ ($A'$ = set of angles such that $x \neq 0$)

cotangent: $A'' \to R$, with $\cot \theta = \dfrac{x}{y}$ ($A''$ = set of angles such that $y \neq 0$)

secant: $A' \to R$, with $\sec \theta = \dfrac{r}{x}$

cosecant: $A'' \to R$, with $\csc \theta = \dfrac{r}{y}$

Note the exclusion of values of $x$ and $y$ that would call for division by zero, an operation having no meaning. For example tan 90° is not defined, since the terminal side of a 90° angle in standard position lies along the positive half of the $y$ axis and hence $P(x, y)$ on any point of it would be a point $(0, y)$; thus $x = 0$ there.

In order to illustrate how each of the above six functions respectively assigns exactly one number to each angle, we now examine a graphical method of obtaining approximate values of these numbers for any given angle $\theta$. Consider Figure 2.5. A pair of rectangular axes is set up as shown,

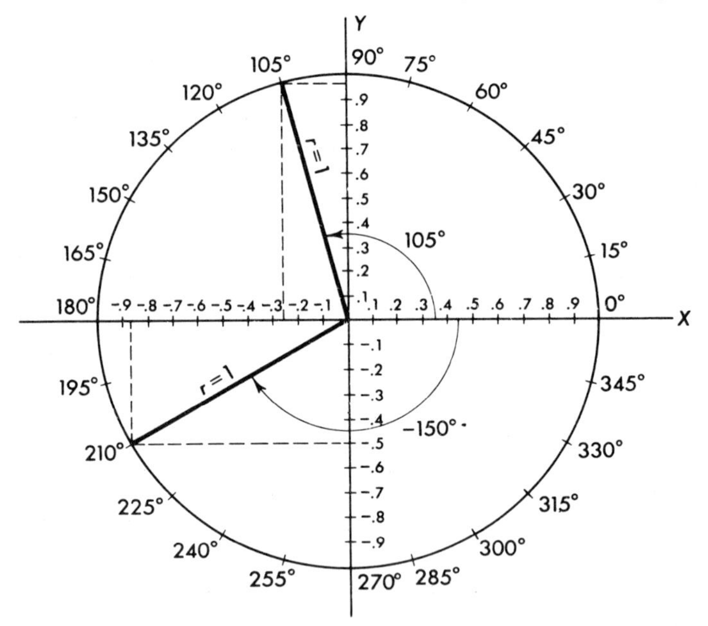


Fig. 2.5 Values of the trigonometric functions from a graph

the unit of distance being chosen large enough to locate tenths of a unit. To facilitate the location of any angle in standard position, arcs of 15-degree intervals from 0° to 360° are marked on a circle of radius 1. The trigonometric functions of any angle, say 105°, are found by locating it on the drawing and choosing the point $(x, y)$ so that the distance $r = 1$; that is, the point $(x, y)$

is on the unit circle. By inspection it is seen that

$$x = -0.26 \quad \text{and} \quad y = 0.97 \text{ (approximately)}$$

Hence, from the previous definitions, approximate values of the trigonometric functions of 105° are*

$$\sin 105° = \frac{y}{r} \approx \frac{0.97}{1} = 0.97$$

$$\cos 105° = \frac{x}{r} \approx \frac{-0.26}{1} = -0.26$$

$$\tan 105° = \frac{y}{x} \approx \frac{0.97}{-0.26} \approx -3.7$$

$$\cot 105° = \frac{x}{y} \approx \frac{-0.26}{0.97} \approx -0.27$$

$$\sec 105° = \frac{r}{x} \approx \frac{1}{-0.26} \approx -3.8$$

$$\csc 105° = \frac{r}{y} \approx \frac{1}{0.97} \approx 1.03$$

The student is reminded that the point $(x, y)$ could have been chosen at *any* point on the terminal side of the angle, the distance $r = 1$ being chosen here simply for convenience in computing ratios.

The values of the six trigonometric functions of an angle are signed numbers, being positive or negative depending upon the signs of $x$ and $y$ in the quadrant in which the angle terminates. The fact that an *angle* is positive or negative does not alone determine the sign of its trigonometric functions.

## *EXERCISES*

1. Construct four figures, each showing a positive angle with a terminal side in a different quadrant. Determine the sign of $x$ and $y$ for each case (what is the sign of $r$?) and state the signs of the six trigonometric ratios for each of the four angles.
2. Repeat Exercise 1, using four negative angles with terminal sides in the various quadrants.
3. Verify the statement: "The sign of sin $\theta$ will always be the same as the sign of $y$." Of what other function can the same statement be made?
4. Complete the statement: "The sign of cos $\theta$ will always be the same as . . ." Can the same be said of any other function? If so, which one?
5. An angle of −150° has been constructed in standard position in Figure 2.5. Taking $r = 1$ as shown, find the values of $x$ and $y$ and verify that $\sin(-150°) = -0.5$, and that $\tan(-150°) = 0.58$. Find also the values of the four remaining functions.

---

* The symbol $\approx$ is read "is approximately equal to."

**6.** Verify the following statement: "If $(x, y)$ is chosen on the terminal side of angle $\theta$ so that $r = 1$, then $\sin \theta = y$ and $\cos \theta = x$."

**7.** Using Figure 2.5 and the results of Exercise 6, find, correct to two figures, the following functions:

(**a**) $\cos 210°$
(**b**) $\sin 140°$
(**c**) $\sin(-30°)$
(**d**) $\cos(-60°)$
(**e**) $\sin 255°$
(**f**) $\cos 120°$
(**g**) $\cos(-225°)$
(**h**) $\sin(-210°)$

**8.** In which quadrant will an angle satisfying each of the following conditions terminate?

(**a**) $\sin \alpha < 0$, $\cos \alpha < 0$
(**b**) $\sin \beta > 0$, $\tan \beta > 0$
(**c**) $\cos \theta > 0$, $\tan \theta < 0$
(**d**) $\tan \varphi < 0$, $\cos \varphi < 0$

**9.** Find the value of each of the six trigonometric functions of an angle $\theta$ which in standard position has its terminal side passing through the point

(**a**) $(-3, 4)$
(**b**) $(12, 5)$
(**c**) $(8, -6)$
(**d**) $(-4, -3)$
(**e**) $(-8, -15)$
(**f**) $(-1, 1)$

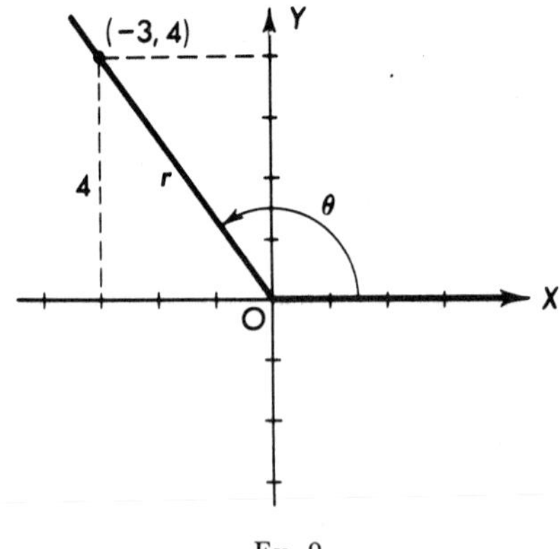


Ex. 9

*Solution of part (a):* We sketch an angle $\theta$ in standard position, as in the figure, with its terminal side passing through the point $(-3, 4)$, selecting the smallest positive angle for convenience. We thus have $x = -3, y = 4$, and $r = \sqrt{x^2 + y^2} = 5$. Hence

$$\sin\theta = \frac{y}{r} = \frac{4}{5}$$

$$\cos\theta = \frac{x}{r} = \frac{-3}{5} = -\frac{3}{5}$$

$$\tan\theta = \frac{y}{x} = \frac{4}{-3} = -\frac{4}{3}$$

$$\cot\theta = \frac{x}{y} = \frac{-3}{4} = -\frac{3}{4}$$

$$\sec\theta = \frac{r}{x} = \frac{5}{-3} = -\frac{5}{3}$$

$$\csc\theta = \frac{r}{y} = \frac{5}{4}$$

**10.** If $\sin\theta = 5/13$ and $\theta$ is in quadrant II, find $\cot\theta$ and $\cos\theta$.
**11.** If $\cos\theta = -4/5$ and $\theta$ is in quadrant III, and $\tan\theta$ and $\sin\theta$.
**12.** Find $\sec\theta$ if $\tan\theta = 2/5$ and $\sin\theta < 0$.
**13.** Find $\tan\theta$ if $\sin\theta = \dfrac{-1}{\sqrt{2}}$ and $\cos\theta < 0$.
**14.** Can $|\sin\theta|$ be greater than 1? can $|\cos\theta|$? Why?
**15.** If $\theta$ is a positive, acute angle ($0 < \theta < 90°$), state which of each of the following pairs is greater and give a reason for your answer.

(**a**) $\sin\theta$ or $\tan\theta$?
(**b**) $\cos\theta$ or $\cot\theta$?
(**c**) $\sec\theta$ or $\tan\theta$?
(**d**) $\csc\theta$ or $\cot\theta$?

**16.** (**a**) Verify the following equalities.

$$\cos 30° = \sin 60° = \sin\,(\text{complement } 30°)$$
$$\cot 30° = \tan 60° = \tan\,(\text{complement } 30°)$$
$$\csc 30° = \sec 60° = \sec\,(\text{complement } 30°)$$

(**b**) Develop three sets of equalities of the form of those in part (a) beginning with $\cos 60°$, $\cot 60°$, and $\csc 60°$, respectively.

**17.** Prove that, for any positive acute angle $\theta$,

$$\cos\theta = \sin\,(\text{complement } \theta)$$
$$\cot\theta = \tan\,(\text{complement } \theta)$$

and

$$\csc\theta = \sec\,(\text{complement } \theta)$$

thus explaining the naming of the functions cosine, cotangent, and cosecant. [*Hint:* If $P(x, y)$ on the terminal side of $\theta$ (in standard position) point is the point $(a, b)$, show that the point $(b, a)$ is on the terminal side of $90° - \theta$ in standard position. Calculate the distance $r$ for each angle, and apply the definitions of the functions.]

**18.** Express each of the following in terms of a function of an angle between $0°$ and $45°$.

(**a**) $\cos 62°$
(**b**) $\sin 85°$
(**c**) $\tan 76°$
(**d**) $\csc 55°$
(**e**) $\cot 78°25'$
(**f**) $\sin 45°\ 30'$
(**g**) $\sec (90° - \theta),\ 0 < \theta < 45°$
(**h**) $\cos (90° - \theta),\ 45° < \theta < 90°$

---

## 2.5 TRIGONOMETRIC FUNCTIONS OF QUADRANTAL ANGLES

An angle in standard position whose terminal side lies along one of the coordinate axes is called a *quadrantal angle*. Angles of $0°$, $90°$, $180°$, $270°$, $360°$ (or, in general, $n \cdot 90°$, where $n = 0, \pm 1, \pm 2$, and so forth) are thus the quadrantal angles. Application of the definitions of the six trigonometric functions of an angle will give the actual values of these functions when they exist. We recall that, if the point $P(x, y)$ on the terminal side of $\theta$ is such that $x = 0$ or that $y = 0$, two of the six trigonometric functions become meaningless and such angles are excluded from the domain. The quadrantal angles are the only angles for which this condition arises. An example will illustrate these statements.

**EXAMPLE**

Find the values of the trigonometric functions of an angle of $90°$.

**Solution:** Figure 2.6 shows an angle of $90°$ in standard position. Then an arbitrary point $P(x, y)$ on the terminal side will be the point $(0, r)$, where, as usual, $r = OP$. We thus have

$$\sin 90° = \frac{y}{r} = \frac{r}{r} = 1$$

$$\cos 90° = \frac{x}{r} = \frac{0}{r} = 0$$

$$\tan 90° = \frac{y}{x} = \frac{r}{0}, \text{ not defined}$$

$$\cot 90° = \frac{x}{y} = \frac{0}{r} = 0$$

$$\sec 90° = \frac{r}{x} = \frac{r}{0}, \text{ not defined}$$

$$\csc 90° = \frac{r}{y} = \frac{r}{r} = 1$$

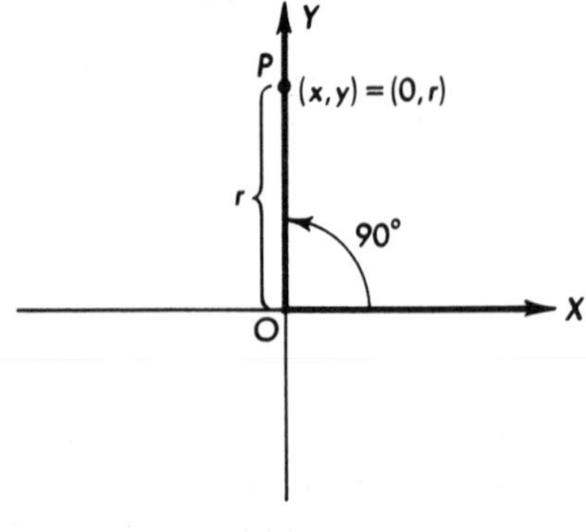


Fig. 2.6 Angle of 90°

Exact values of the functions (when defined) of the other quandrantal angles can be found in the same manner. The student should practice making these evaluations until he can determine the values from a "mental sketch" without hesitation. There is a great difference between ready evaluation by the above process and a rote memorization with little or no understanding.

## *EXERCISES*

1. Division of a number $a$ by a number $b$ may be defined in terms of finding a number $c$ such that $cb = a$. For example $28 \div 4 = 7$ because $7 \times 4 = 28$. Use this idea to show that if $a \neq 0$, then there is no number $c$ that is the result of dividing $a$ by 0, thus showing that division of a non-zero number by 0 is *not defined.*
2. Repeat the approach of Exercise 1 to show that $0 \div 0$ is not defined because *any* proposed quotient checks, and hence there is no single (unique) answer.
3. Which of the six trigonometric functions are undefined for (**a**) 0°; (**b**) 90°; (**c**) 180°; (**d**) 270°; (**e**) 360°?
4. Justify the statement made in the paragraph preceding the Example of this section that the quadrantal angles are the only angles for which any of the functions are not defined.
5. Justify the statement that the domain of the tangent function is the set of all angles except the odd multiples of 90°.
6. In the manner of Exercise 5, give the domain of the cotangent function, the secant function, the cosecant function, and the cosine function.
7. Use the method of the Example above to evaluate the trigonometric functions of the following angles (if the functions exist): (**a**) 0°; (**b**) 180°; (**c**) 270°; (**d**) 360°; (**e**) −270°.

## 2.6 TRIGONOMETRIC FUNCTIONS OF 30°, 45°, AND THEIR MULTIPLES

In Section 2.4 we showed how to approximate values of the trigonometric functions of any angle by graphical means. In Section 2.7 we shall discuss the more accurate approximations of the functions found in tables. Thus far, the quadrantal angles are the only angles for which we have succeeded in obtaining *exact values* of the functions. We now show how to find exact values of the functions of 30°, 45°, and their multiples; we cannot do this for angles in general. Special geometric properties of these angles enable us to make the evaluations. We shall find a knowledge of these exact values most useful.

We first obtain values of the functions of 45°. The geometric property utilized here is that, if a point $P(x, y)$ is chosen on the terminal side of an angle of 45° in standard position, then $x = y$. Since values of the trigonometric functions are independent of the location of $P$ along the terminal side, we choose it so that $x = y = 1$, and hence $r = \sqrt{2}$ (see Figure 2.7). Then

$$\sin 45° = \frac{y}{r} = \frac{1}{\sqrt{2}}$$

$$\cos 45° = \frac{x}{r} = \frac{1}{\sqrt{2}}$$

$$\tan 45° = \frac{y}{x} = \frac{1}{1} = 1$$

$$\cot 45° = \frac{x}{y} = \frac{1}{1} = 1$$

$$\sec 45° = \frac{r}{x} = \frac{\sqrt{2}}{1} = \sqrt{2}$$

$$\csc 45° = \frac{r}{y} = \sqrt{2}$$

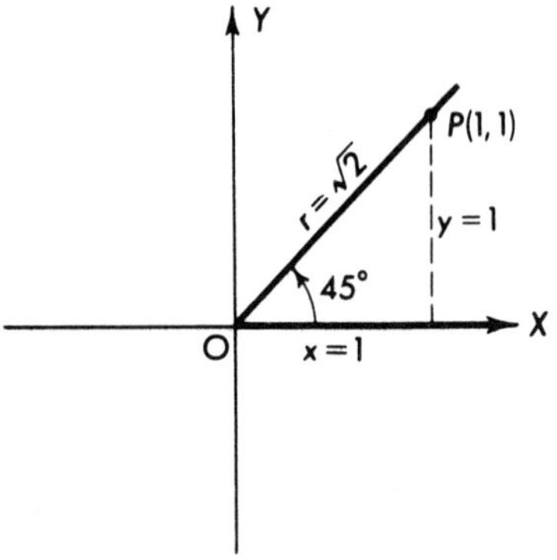

Fig. 2.7 Angle of 45°

Functions of the *odd* multiples of 45° may be evaluated by using the fact that $|y| = |x|$ for the point $P(x, y)$ on the terminal side of such angles. (What about the *even* multiples of 45°?) For example, to evaluate the functions of $-135°$ we sketch the angle in standard position and choose $(-1, -1)$ as the point $P$ on the terminal side (Figure 2.8). Again $r = \sqrt{2}$. Thus

$$\sin(-135°) = \frac{y}{r} = \frac{-1}{\sqrt{2}} = -\frac{1}{\sqrt{2}}$$

$$\cos(-135°) = \frac{x}{r} = \frac{-1}{\sqrt{2}} = -\frac{1}{\sqrt{2}}$$

$$\tan(-135°) = \frac{y}{x} = \frac{-1}{-1} = 1$$

$$\cot(-135°) = \frac{x}{y} = \frac{-1}{-1} = 1$$

$$\sec(-135°) = \frac{r}{x} = \frac{\sqrt{2}}{-1} = -\sqrt{2}$$

$$\csc(-135°) = \frac{r}{y} = \frac{\sqrt{2}}{-1} = -\sqrt{2}$$

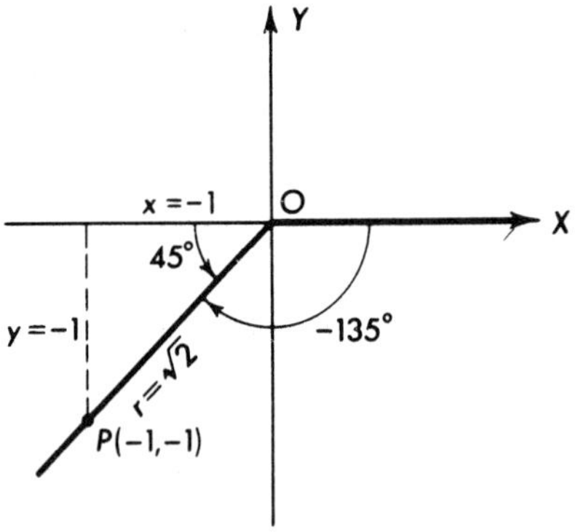

Fig. 2.8 Angle of $-135°$

The property that enables us to evaluate the functions of an angle of 60° is that, if a 60° angle is constructed in standard position, then, for the arbitrary point $P$, we have $x = \frac{1}{2}r$. (This is evident if we remember that each angle of an equilateral triangle is 60°, and that the altitude from one vertex of an equilateral triangle bisects the opposite side.) Hence, if we choose $P$ so that $r = 2$, we have $x = 1$, and then $y = \sqrt{2^2 - 1} = \sqrt{3}$ (Figure 2.9).

Thus

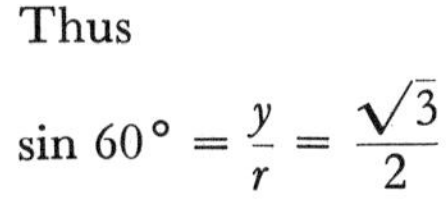

$$\sin 60° = \frac{y}{r} = \frac{\sqrt{3}}{2}$$

$$\cos 60° = \frac{x}{r} = \frac{1}{2}$$

$$\tan 60° = \frac{y}{x} = \frac{\sqrt{3}}{1} = \sqrt{3}$$

$$\cot 60° = \frac{x}{y} = \frac{1}{\sqrt{3}}$$

$$\sec 60° = \frac{r}{x} = \frac{2}{1} = 2$$

$$\csc 60° = \frac{r}{y} = \frac{2}{\sqrt{3}}$$

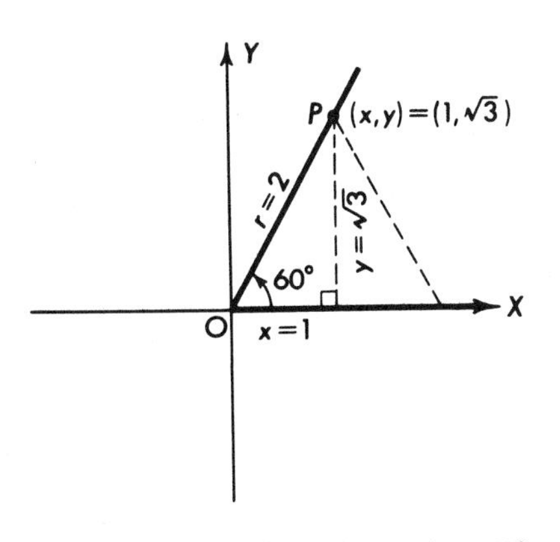


Fig. 2.9 Angle of 60°

Similarly, if $P$ is chosen on the terminal side of an angle of 30° in standard position so that $r = 2$, we have $x = \sqrt{3}$, $y = 1$, as in Figure 2.10. Again applying the definitions of the functions and simplifying, we have

$$\sin 30° = \frac{1}{2}$$

$$\cos 30° = \frac{\sqrt{3}}{2}$$

$$\tan 30° = \frac{1}{\sqrt{3}}$$

$$\cot 30° = \sqrt{3}$$

$$\sec 30° = \frac{2}{\sqrt{3}}$$

$$\csc 30° = 2$$

Fig. 2.10 Angle of 30°

Exact values of the functions of multiples of 30° and 60° that are not quadrantal angles may be evaluated in the manner of the example for the functions of −135°. In Figure 2.11, sketches for some of these angles are

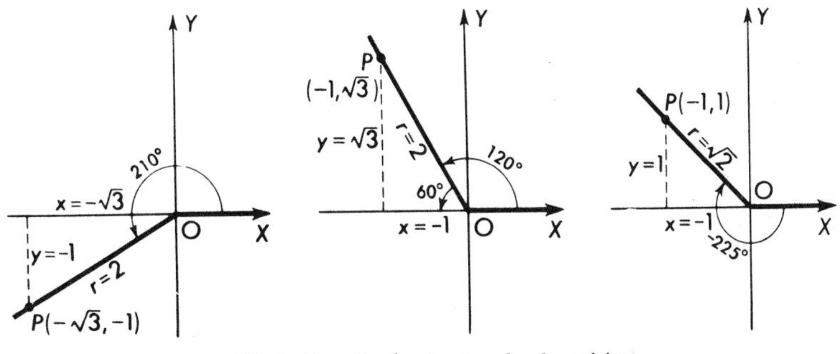


Fig. 2.11 Angles in standard position

given. They should be suggestive in the solution of many of the exercises below.

As with the quadrantal angles, it is urged that the student practice creating mental pictures of the figures used to evaluate the functions of 30°, 45°, and 60°. The ability to give these values quickly will be indispensable in many exercises through the book. Try this on the following exercises. Verify the following:

## *EXERCISES*

Verify the following:

1. $\sin 30° + \cos 60° = 1$
2. $\sin 60° = 2 \sin 30° \cos 30°$
3. $\cos 60° = (\cos 30°)^2 - (\sin 30°)^2$
4. $\sin 90° = 2 \sin 45° \cos 45°$
5. $\cos 90° = 1 - 2(\sin 45°)^2$
6. $\cos 90° = \cos 60° \cos 30° - \sin 60° \sin 30°$
7. $\sin 120° = \sin 90° \cos 30° + \cos 90° \sin 30°$
8. $\tan 30° \cot 60° + \tan 45° = \dfrac{\sec 30°}{\cos 30°}$
9. $\cos 45° = \sqrt{\dfrac{1 + \cos 90°}{2}}$
10. Evaluate the six trigonometric functions of each of the following: (**a**) 150°; (**b**) 240°; (**c**) 330°; (**d**) −300°
11. Verify the following by evaluating each member:
    (**a**) $\sin 120° = \sin 60°$
    (**b**) $\cos 330° = \cos 30°$
    (**c**) $\tan 225° = \tan 45°$
    (**d**) $\sin 240° = -\cos 30°$
    (**e**) $\cos 210° = -\sin 60°$
    (**f**) $\tan 120° = -\cot 30°$
    (**g**) $\sin 300° = -\cos 30°$
12. Find the value of each of the following:
    (**a**) $(\sin 30°)^2 + (\cos 30°)^2 + (\tan 30°)^2 - (\sec 30°)^2$
    (**b**) $\sin 240° \cos 30° + \cos 210° \sin 300°$
    (**c**) $(\sin 120°)^2 + (\cos 120°)^2$
    (**d**) $(\tan 240°)^2 - (\sec 240°)^2$
    (**e**) $2 \sin 120° \cos 120°$
13. For what values of $\theta$, if any, is each of the following undefined: (**a**) $\sin \theta$; (**b**) $\cos \theta$; (**c**) $\tan \theta$; (**d**) $\cot \theta$; (**e**) $\sec \theta$; (**f**) $\csc \theta$.
14. *A mettle tester.* Find the exact values of sin 36° and cos 36°.

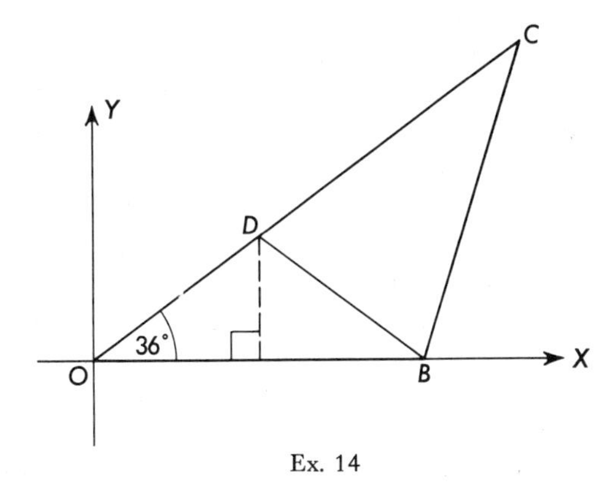


Ex. 14

*Suggestion:* With a 36° angle in standard position as in the figure, construct isosceles triangle $OBC$ as shown with $OB = 2$ and $BC = 2$. Find point $D$ on $OC$ so that $CD = 2$. Using the fact that the sum of the angles of any triangle is 180°, find the measures of all the angles that you can. With the aid of the similar triangles you discover, and the Pythagorean relation, you should be able to determine the coordinates of the point $D$ and the distance $r = OD$.

## 2.7 VALUES OF TRIGONOMETRIC FUNCTIONS FROM TABLES

In the preceding section we computed values of the trigonometric functions of special angles. Many of the applications of trigonometry depend upon knowing the values of the functions of *any* angle. The calculation of these, in general, is beyond the scope of this book. By use of higher mathematics, values of these functions can be approximated to any desired degree of accuracy. Table III in the back of this book gives, correct to four places, values of the trigonometric functions of angles from 0° to 90° by 10-min. intervals. In the next section we shall show that the values of the functions of any other angle can be expressed in terms of the values for the acute angles found in the table.

The names of the functions given at the *top* of each section of the table refer to the angle in the *left* column. Thus, $\sin 24° 40' = 0.4173$, $\tan 7° 20' = 0.1287$, and $\cot 44° = 1.0355$. Angles up to 45° can thus be located. The names of the functions given at the *bottom* of each section refer to the angles on the *right*. For example, $\cos 67° 20' = 0.3854$ and $\sin 55° 30' = 0.8241$. Note how the space occupied by the table of trigonometric functions has been halved by the use of the relations $\cos \theta = \sin (90° - \theta)$, $\sin \theta = \cos (90° - \theta)$, $\tan \theta = \cot (90° - \theta)$, and so forth (compare Exercises 16–18, Section 2.4).

In order to approximate values of the trigonometric functions of angles given to the nearest minute, but not occurring in the table (for example, $\sin 38° 25'$), we use linear interpolation. We make the assumption that the variation in a trigonometric function of an angle is proportional to the variation in the angle. This is equivalent to the assumption that the graphs of these functions are straight lines. In general, this is not a justifiable assumption. However, *if the variation of the angle is over an interval of no more than 10 min*, it can be proved that the deviation of the true value of the sine function or the cosine function from the value found by linear interpolation is not detectable in the fourth significant figure. Thus we are sure of a reliable approximation to the sine or cosine function when we interpolate over any of the 10-min intervals found in the table. Similar statements apply to each of the other trigonometric functions, tangent, cotangent, secant, and cosecant, *provided the angles are not close* to the values for which the function is undefined. For example, interpolation is unreliable for $\tan \theta$ if $80° < \theta < 100°$. Interpolation is explained further by means of examples.

### EXAMPLE 1

Find $\sin 38° 25'$.

**Solution:** The value of $\sin 38° 25'$ must lie between $\sin 38° 20'$ and $\sin 38° 30'$. Since $38° 25'$ is halfway between $38° 20'$ and $38° 30'$, we assume that its sine is approximately halfway between $\sin 38° 20'$ and $\sin 38° 30'$. Hence

$$\begin{aligned} \sin 38° 25' &\approx \sin 38° 20' + 0.5 (\sin 38° 30' - \sin 38° 20') \quad \text{(subtract mentally)} \\ &\approx 0.6202 + 0.5(0.0023) \\ &\approx 0.6202 + 0.0012 \quad \text{(to nearest ten-thousandth)} \\ &= 0.6214 \end{aligned}$$

(Compare this with Exercise 5, Section 1.5.)

### EXAMPLE 2

Find $\tan 73° 7'$.

**Solution:** As above,

$$\begin{aligned} \tan 73° 7' &\approx \tan 73° 0' + 0.7 (\tan 73° 10' - \tan 73° 0') \\ &\approx 3.271 + 0.7(0.034) \\ &\approx 3.271 + 0.024 \\ &= 3.295 \end{aligned}$$

### ⇨ EXAMPLE 3

Find $\cos 16° 43'$.

**Solution:**

$$\begin{aligned} \cos 16° 43' &\approx \cos 16° 40' + 0.3\,(\cos 16° 50' - \cos 16° 40') \\ &\approx 0.9580 + 0.3(0.9572 - 0.9580) \\ &= 0.9580 + 0.3(-0.0008) \\ &\approx 0.9580 - 0.0002 \\ &= 0.9578 \end{aligned}$$

If the value of one of the trigonometric functions of an angle is known and the size of the angle to the nearest minute is to be determined, apply an interpolation method to determine an angle whose given function value does not appear in the table. The following examples make the method clear.

### ⇨ EXAMPLE 4

If $\tan \alpha = 0.6148$, find $\alpha$.

**Solution:** In Table III, 0.6148 does not appear in the tangent column for any angle. However, it is found to lie between $\tan 31° 30' = 0.6128$ and $\tan 31° 40' = 0.6168$; in fact, it is readily seen to be approximately midway between them. Hence we assume that $\alpha$ is approximately midway between $31° 30'$ and $31° 40'$. Hence

$$\alpha \approx 31° 35'$$

### ⇨ EXAMPLE 5

If $\sin \alpha = 0.5040$, find $\alpha$.

**Solution:** From Table III note that $\sin 30° 10' = 0.5025$ and that $\sin 30° 20' = 0.5050$. Hence $\alpha = 30° 10' +$ (some fraction of $10'$). By inspection, 0.5040 is 15/25 or 0.6 of "the way" between 0.5025 and 0.5050. Hence

$$\alpha \approx 30° 10' + 0.6(10') = 30° 16'$$

### ⇨ EXAMPLE 6

If $\sin \alpha = 0.4419$, find $\alpha$.

**Solution:** By inspection,

$$\begin{aligned} \alpha &\approx 26° 10' + \frac{9}{26}(10') \\ &\approx 26° 10' + 3' \\ &= 26° 13' \end{aligned}$$

**EXAMPLE 7**

If $\cos \beta = 0.4059$, find $\beta$.

**Solution:** By inspection,

$$\beta \approx 66° \, 0' + \frac{8}{26}(10') \approx 66° \, 3'$$

## *EXERCISES*

Verify that the following values of trigonometric functions are correct to the accuracy given.

1. $\sin 19°20' = .3311$
2. $\cos 61°44' = .4736$
3. $\tan 12°9' = .2153$
4. $\sin 29°32' = .4929$
5. $\cot 36° = 1.3764$
6. $\cos 37°18' = .7955$
7. $\tan 81°37' = 6.786$
8. $\sin 48°42' = .7513$
9. $\cot 81°13' = .1545$
10. $\cos 18°4' = 0.9507$
11. $\tan 65°44' = 2.2182$
12. $\cot 84°52' = .08983$
13. $\sec 29°17' = 1.1465$
14. $\csc 64°42' = 1.1061$
15. $\sec 55°33' = 1.7678$
16. $\csc 38°14' = 1.6159$

Find correct to the nearest minute the acute angles for which values of trigonometric functions are given below.

17. $\sin \alpha = 0.4025$
18. $\cos \theta = 0.7570$
19. $\tan \beta = 1.0650$
20. $\cot \alpha = 1.1760$
21. $\sec \theta = 2.0830$
22. $\csc \beta = 2.2050$
23. $\cos \alpha = 0.5530$
24. $\sin \theta = 0.8815$
25. $\cot \alpha = 0.4030$
26. $\tan \beta = 0.6680$
27. $\csc \theta = 1.3950$
28. $\sec \phi = 1.1860$
29. $\sin \beta = 0.4592$
30. $\cos \delta = 0.6947$

## 2.8 FINDING THE VALUES OF THE FUNCTIONS OF ANY ANGLE. REDUCTION TO FUNCTIONS OF AN ACUTE ANGLE

Table III, functions of angles from 0° to 90°, can be used to determine the values of the functions of *any* angle. Assuming that $h$ and $k$ are positive numbers, plot the points $P(h, k)$, $Q(-h, k)$, $R(-h, -k)$, and $S(h, -k)$ on a coordinate system as shown in Figure 2.12. The absolute values of the

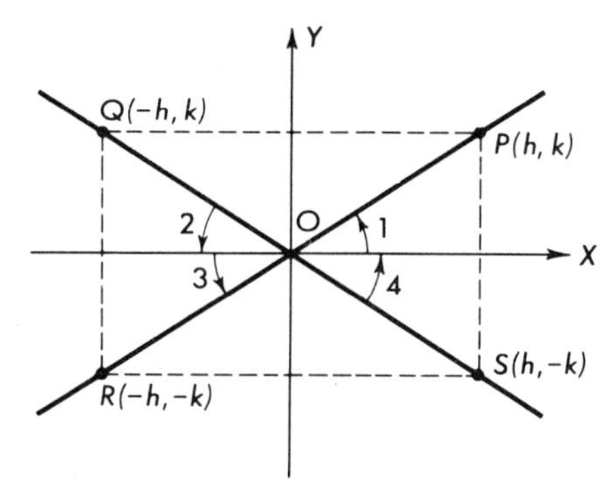


Fig. 2.12 Relation to functions of acute angle

corresponding coordinates of these four points are equal, since $|-h| = h$ and $|-k| = k$. Also, the absolute distance $r$ of each of these points from the origin is the same, namely, $r = \sqrt{h^2 + k^2}$. From this and from the definitions of the trigonometric functions, it follows that the angles $XOP$, $XOQ$, $XOR$, $XOS$ and any angles in standard position co-terminal with them *have trigonometric functions of the same absolute value.* Furthermore, angles 1, 2, 3, and 4 of the figure are equal. Thus any trigonometric function of $\angle XOQ$ is *numerically equal* to the value of the same function of $\angle 1 = \angle 2$. For example, $|\cos \angle XOQ| = \cos \angle 2$. Again, any function of $\angle XOR$ is equal in absolute value to the same function of $\angle 3$. Likewise, $|\tan \angle XOS| = \tan \angle 4$, and so forth. Of course, the *sign* of each function of one of these angles is determined by the quadrant in which the angle terminates.

In summary, *any function of any angle may be reduced to plus or minus the same function of an angle in the interval from 0° to 90° by the following three steps:*

(**a**) *Sketch the angle in standard position.* (A mental sketch may suffice.)

(**b**) *Determine the sign of the desired trigonometric function from the signs of x and y at a point on its terminal side.*

(**c**) *Use the fact that the numerical value of the desired function is equal to that of smallest non-negative angle formed by the X axis and the terminal side of the given angle.* (The latter angle will be called the *reference angle* corresponding to the given angle.)

**EXAMPLE 1**

Find cos 217° 24′.

**Solution (by the above steps):**

**A** A sketch of the angle in standard position shows that it terminates in the third quadrant (Figure 2.13a).

**B** $\cos 217°\ 24' = x/r$, which is *negative* (in quadrant III).
**C** The reference angle is $217°\ 24' - 180° = 37°\ 24'$. Hence

$$\cos 217°\ 24' = -\cos 37°\ 24' \approx -0.7944$$

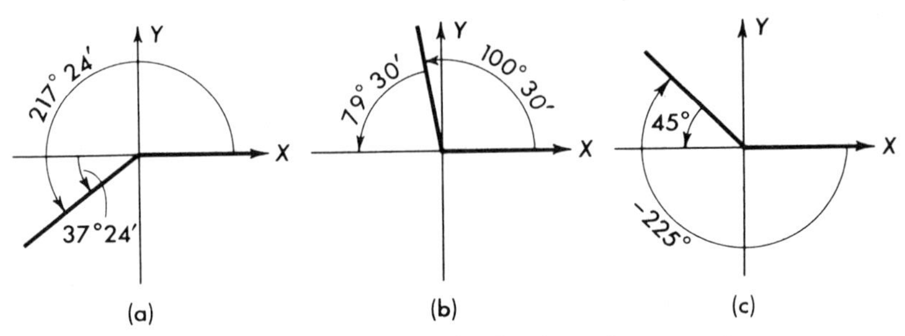


Fig. 2.13 Sketches of angles showing reference angles

**EXAMPLE 2**

Find $\tan 100°\ 30'$.

**Solution (by the above steps):** The angle terminates in the second quadrant (Figure 2.13b). The sign of $\tan 100°\ 30'$ is negative in this quadrant. The reference angle is $180° - 100°\ 30' = 79°\ 30'$. Hence

$$\tan 100°\ 30' = -\tan 79°\ 30' \approx -5.396$$

**EXAMPLE 3**

Without tables, find the exact value of $\csc(-225°)$.

**Solution:** From the sketch (Figure 2.13c), it is seen that $-225°$ terminates in quadrant II, so that the sign of $\csc(-225°)$ is positive. The reference angle is 45°. Hence

$$\csc(-225)° = \csc 45° = \sqrt{2}$$

**EXAMPLE 4**

Given $\sin\theta = -0.4067$ and $\cos\theta > 0$, find $\theta$.

**Solution:** The only quadrant in which $\sin\theta < 0$ and $\cos\theta > 0$ is quadrant IV. From Table III, $\sin 24° \approx 0.4067$. Hence the reference angle is 24°. This and the fact that $\theta$ is in quadrant IV give

$$\theta \approx 360° - 24° = 336°$$

(Of course, $\theta$ may be any angle co-terminal with 336°.)

## *EXERCISES*

1. Express each of the following in terms of the same function of a positive acute angle and evaluate without tables.

   (a) $\sin 120°$
   (b) $\cos 225°$
   (c) $\tan 300°$
   (d) $\cot 150°$
   (e) $\sec 330°$
   (f) $\csc 210°$
   (g) $\sin(-30°)$
   (h) $\cos(-120°)$
   (i) $\tan(-60°)$
   (j) $\cot(-210°)$
   (k) $\sec(-300°)$
   (l) $\csc(-225°)$

2. Express each of the following in terms of the same function of a positive acute angle. Do not evaluate.

   (a) $\cos 200°15'$
   (b) $\sin 95°52'$
   (c) $\tan 310°37'$
   (d) $\cot 101°14'$
   (e) $\sec 170°$
   (f) $\csc 200°$
   (g) $\tan(-100°)$
   (h) $\cot(-214°)$
   (i) $\sin(-300°18')$
   (j) $\cos(-44°38')$
   (k) $\csc(-81°12')$
   (l) $\sec(-245°43')$

3. In the interval $0 \leqq \theta < 360°$, find $\theta$ determined by each of the following:

   (a) $\sin\theta = 0.5$, given that $\tan\theta < 0$
   (b) $\cos\theta = 0.5$, " " $\csc\theta < 0$
   (c) $\tan\theta = -1$, " " $\sin\theta > 0$
   (d) $\cot\theta = \sqrt{3}$, " " $\cos\theta < 0$
   (e) $\sec\theta = 2$, " " $\sin\theta < 0$
   (f) $\csc\theta = -\sqrt{2}$, " " $\cos\theta > 0$
   (g) $\cos\theta = -0.7835$, " " $\csc\theta > 0$
   (h) $\sin\theta = 0.8455$, " " $\cos\theta < 0$

## *MISCELLANEOUS EXERCISES/CHAPTER 2*

**1.** Determine whether each of the following statements is true or false.

(**a**) $(\sin 120°)^2 + (\cos 120°)^2 = 1$
(**b**) $\sin 300° \cos 300° = \sin 600°$
(**c**) $\tan 150° = \dfrac{\cos 300°}{\sin 300°}$
(**d**) $\dfrac{2 + \cos 300° - \sec 60°}{1 - \sin 150° + \tan 135°} = -1$
(**e**) $\dfrac{1 + \cot 30° - \tan 60°}{2 + \sin 30° + \cos 240°} = 2$
(**f**) $\dfrac{\cos 420° - \cot 570° + \sec 480°}{\csc (-330°) + \sin 530° - \tan 405°} = \sqrt{3}$
(**g**) For every angle $\theta$, $\cos (-\theta) = -\cos \theta$.
(**h**) If $45° < \theta < 90°$, then $\sin \theta > \cos \theta$.
(**i**) Any trigonometric function of any angle is never greater than the same function of the reference angle.
(**j**) The only angles that are ever excluded from the domain of any of the trigonometric functions are quadrantal angles.

**2.** In using interpolation to approximate the values of trigonometric functions of positive acute angles, you probably noticed that you added or subtracted your correction according as you were approximating a function *without* or *with* the prefix "co" in its name. Why is this?

**3.** Check the statement of Exercise 2 for approximating the six functions of $23°43'$.

**4.** For each of the following, find the smallest positive angle $\theta$ satisfying the given conditions.

(**a**) $\sin \theta = \dfrac{\sqrt{3}}{2}$, $\cos \theta = -\frac{1}{2}$
(**b**) $\tan \theta = 1$, $\sin \theta < 0$
(**c**) $\cos \theta = -\frac{1}{2}$, $\theta$ is in quadrant II
(**d**) $\tan \theta = -\sqrt{3}$, and $\theta$ is an angle of a triangle.
(**e**) $\tan \theta = \cot \theta$, and both are negative.
(**f**) The terminal side of $\theta$ passes through the point $(-\sqrt{3}, -1)$.

**5.** Explain how we could "get by" if we had tables of the trigonometric functions that included only angles in the interval $[0, 45°]$.

**6.** Consider the definining equation for the sine function, $\sin \theta = y/r$. Give a condition that is necessary in order to have $\sin \theta = 0$.

Give a general description of the set of angles whose sine is zero.

7. Repeat Exercise 6 for cos $\theta$.
8. Show that cos $\theta = 0$ if and only if cot $\theta = 0$. (Note: "if and only if" statements are actually two statements combined. The "if" part here is the statement "cos $\theta = 0$ if cot $\theta = 0$," which is equivalent to "If cot $\theta = 0$ then cos $\theta = 0$." The "only if" part is the statement "cos $\theta = 0$ only if cot $\theta = 0$," which is equivalent to the statement "If cos $\theta = 0$ then cot $\theta = 0$." This means that to prove the given statement you must prove a statement and its converse.)
9. Show that sin $\theta = 0$ if and only if tan $\theta = 0$.
10. Formulate a statement identifying all values of $\theta$ for which cos $\theta = 1$; for which cos $\theta = -1$.
11. Repeat Exercise 10 for sin $\theta$; for tan $\theta$.
12. Compare the value of cos $(-60°)$ with that of cos $60°$; of cos $(-150°)$ with that of cos $150°$, of cos $(-225°)$ with that of cos $225°$; of cos $(-300°)$ with that of cos $300°$. What general conclusion do you think is likely? (We shall return to this matter in Chapter 6.)
13. Repeat Exercise 12 for *sine* instead of *cosine*.
14. Each trigonometric function is positive for angles in two quadrants and is negative for angles in the remaining two quadrants. In which two quadrants is sin $\theta$ positive? Give the corresponding information for each of the other five functions.

# CHAPTER 3. RADIAN MEASURE OF ANGLES. TRIGONOMETRIC FUNCTIONS OF A NUMBER

## 3.1 RADIAN MEASURE

Thus far we have used the degree as the fundamental unit of angle measure. We now introduce another important unit, the *radian.** An angle of one radian, if placed with its vertex at the center of a circle, intercepts an arc whose length is equal to that of the radius of the circle. Thus, if in Figure 3.1

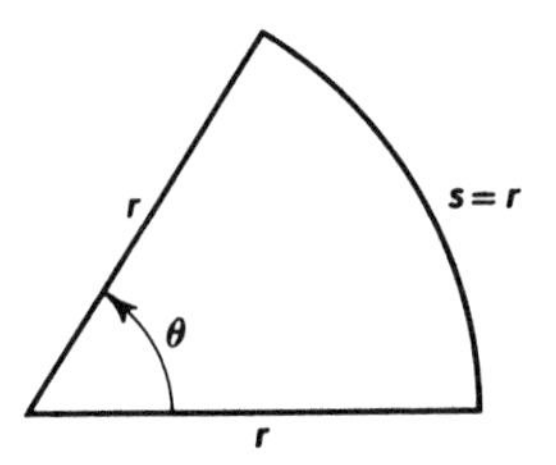


Fig. 3.1 Angle of measure 1 rdn

the length of the circular arc $s$ is equal to the radius $r$, then the central angle $\theta$ is an angle of one radian. This simple relation of "unit of arc length to unit of angle measure" is as important in mathematics as is the relation "one cubic centimeter of water weighs one gram" in chemistry or physics. Many of the formulas of calculus are simplified if angles are measured in radians.

The abbreviation for both *radian* and *radians* is *rdn*. *If no unit of angle measure is specified, it will be understood that the angle is given in radians.* Thus, if $\theta$ is an angle, the statement $\theta = 3$ means $\theta$ is an angle of 3 rdn.

The number $\pi$ plays an important role in radian measure. By definition,

$$\pi = \frac{\text{circumference of a circle}}{\text{diameter of the circle}}$$

$$= \frac{c}{2r}$$

* Other units of angle measure are: the *right angle;* the *revolution;* the *mil*, 1/6400 of a revolution.

$\pi$ is a real constant the value of which to nine significant figures is 3.14159265. In the following section we develop the role of $\pi$ in radian measure of angles.

## 3.2 CHANGE OF UNIT OF ANGLE MEASURE

From the definition of $\pi$, if $c$ is the circumference and $r$ the radius of a circle,

$$c = 2\pi r$$

Arcs of length, $c$ and $r$ correspond to central angles of 1 revolution and 1 rdn, respectively. Hence

$$1 \text{ revolution} = 2\pi \text{ rdn}$$

And, since

$$1 \text{ revolution} = 360^\circ$$

therefore

$$360^\circ = 2\pi \text{ rdn}$$

or

$$180^\circ = \pi \text{ rdn} \qquad \textbf{(3.1)}$$

Equation (3.1) is a basis for changing the measure of a given angle from degrees to radians, or vice versa. If an angle in degrees bears an obvious, simple ratio to 180°, the change to radian measure can be made with the aid of (3.1) by inspection. For example, 90° is ½ of 180°, so that by (3.1)

$$90^\circ = \frac{\pi}{2} \text{ rdn}$$

Again, 60° = ⅓ of 180°, giving

$$60^\circ = \frac{\pi}{3} \text{ rdn}$$

Or, as an example of changing from radian to degree measure by inspection,

$$\frac{3\pi}{2} \text{ rdn} = \frac{3}{2}(180^\circ) = 270^\circ$$

If the fractional relation is not obvious, the change can be made by simple proportion, as illustrated below by examples.

**EXAMPLE 1**

Express 105° in radian measure.

**Solution:** Let

$$x \text{ rdn} = 105^\circ$$

By (3.1),

$$\pi \text{ rdn} = 180^\circ$$

Hence the proportion

$$\frac{x}{\pi} = \frac{105}{180}$$

that yields

$$x = \frac{7\pi}{12} \text{ rdn}$$

(*Note:* By using the approximate relation $\pi = 22/7$, we may express our answer in the form

$$x \approx \frac{11}{6} \text{ rdn}$$

or, using the better approximation, $\pi = 3.14159$, we could have written

$$x = 2.4435 \text{ rdn},$$

a more accurate approximation. However, the form $\frac{7\pi}{12}$ has the advantages of preciseness and an obvious relation to 180°, so we prefer it unless further calculations require a numerical approximation.)

**EXAMPLE 2**

Express 2 rdn in degree measure.

**Solution:** Let

$$x^\circ = 2 \text{ rdn}$$

By (2.1),

$$180^\circ = \pi \text{ rdn}$$

Hence

$$\frac{x}{180} = \frac{2}{\pi}$$

from which

$$x = \frac{360}{\pi}$$

(*Note:* Tables giving the equivalent degree and radian measures of various angles are available. They should be consulted if continued need arises. Table V in this book may help. See also Exercise 6 of the Miscellaneous Exercises at the end of this chapter.)

## *EXERCISES*

**1.** Express in degree measure angles of the following radian measure:

(**a**) $\frac{2\pi}{3}$ (**d**) 2 (**g**) $-\frac{3\pi}{2}$

(**b**) $\frac{\pi}{3}$ (**e**) 0.84 (**h**) $\frac{11\pi}{6}$

(**c**) 0.1 (**f**) $-\frac{\pi}{4}$ (**i**) $\frac{5\pi}{4}$

**2.** Find the radian measure of angles of the following degree measure:

(**a**) $30°$ (**b**) $-270°$ (**c**) $100°$ (**d**) $2°$
(**e**) $-120°$ (**f**) $1080°$ (**g**) $18°$ (**h**) $45'$
(**i**) $72°$ (**j**) $22°30'$ (**k**) $135°$ (**l**) $210°$

**3.** Find the exact value of each of the following without use of tables:

(**a**) $\sin \frac{\pi}{6}$ (**b**) $\cot \pi$
(**c**) $\cos \left(\frac{-2\pi}{3}\right)$ (**d**) $\sec \left(\frac{-\pi}{3}\right)$
(**e**) $\tan \frac{3}{4}\pi$ (**f**) $\cos \frac{4\pi}{3}$

**4.** We have remarked that radian measure of angles is employed in calculus. Values of the functions of positive acute angles with radian measure are given in Table V. It is important to be able to use such a table for evaluations and not to resort to converting to degree measure. For example, to evaluate cos 2, we employ the three steps of Section 2.8 as follows: (1) Note that an angle of 2 rdn terminates in quadrant II, so that its cosine is negative. (**2**) The reference angle is $\pi - 2 \approx 3.14 - 2 = 1.14$. (**3**) By Table V, $\cos 1.14 \approx 0.4176$, so that $\cos 2 \approx -0.4176$. In short, $\cos 2 = -\cos (\pi - 2) \approx -\cos 1.14 \approx -0.4176$.

In the above manner, approximate the value of each of the following:

(**a**) $\tan 2$ (**b**) $\cos 2.3\pi$ (**c**) $\sin 8$
(**d**) $\sin 4$ (**e**) $\sin \left(-\frac{10\pi}{7}\right)$ (**f**) $\tan (-3)$
(**g**) $\sin 1$ (**h**) $\cos 1.3$ (**i**) $\cos \sqrt{2}$

**5.** Determine, in radian measure, the angles between 0 and $2\pi$ for which each of the following is a true statement. Use Table V if a table is needed.

(**a**) $\cos \theta = \frac{1}{2}$
(**b**) $\tan \theta = -1$
(**c**) $\sin \theta = -\frac{\sqrt{3}}{2}$
(**d**) $\sin \theta = 0.5810$
(**e**) $\tan \theta = 0.2664$
(**f**) $\cos \theta = -0.1633$

## 3.3 THE CENTRAL ANGLE AND THE INTERCEPTED ARC

From the definition of a radian it follows that, in a given circle, a central angle of one radian intercepts an arc equal in length to the radius. (See

Figure 3.2.) Hence, by proportion, a central angle of $\theta$ radians intercepts an arc of length $\theta$ times the radius. Otherwise stated,

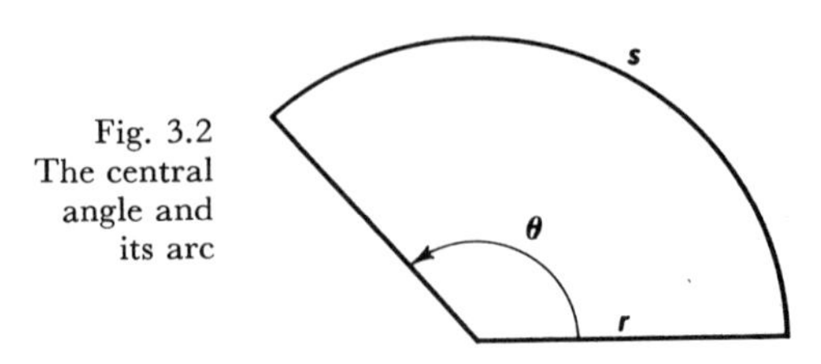


Fig. 3.2 The central angle and its arc

(length of arc) = (measure of subtended angle, in radiaus) $\times$ (length of radius),

or, symbolically,

$$s = \theta r \quad (\theta \text{ in radians}) \tag{3.2}$$

From this formula, any one of the three quantities, $s$, $\theta$, or $r$, can be found if the other two are known.

**EXAMPLE**

What central angle will intercept an arc of length 12 in. on a circle 60 in. in diameter?

**Solution:** From (3.2),

$$r\theta = s$$

Substituting $r = 30$ and $s = 12$ gives

$$30\theta = 12$$

from which

$$\theta = \tfrac{2}{5} \text{ rdn}$$

## *EXERCISES*

**1.** In each of the following, two of the three quantities $r$, $\theta$, and $s$ are given. Find the missing quantity. Other literal numbers represent fixed positive numbers.

(**a**) $r = 5$ in, $\theta = \dfrac{2\pi}{3}$

(**b**) $\theta = 2.3$, $s = 4$ in.

(**c**) $r = 2$ ft, $s = 3$ in.

(**d**) $s = $ a ft, $r = 6$ ft

(**e**) $s = 2m$ ft, $r = \tfrac{1}{2}m$ ft

(**f**) $\theta = k°$, $r = n$ in.

**2.** On a circle, the length of the arc subtending a central angle is equal to the length of the radius. Find the angle in radians; in degrees.

3. Find the distance traveled by the tip of a 6 in. hour hand of a clock from noon to 4 pm.
4. Find the distance along $1^\circ$ of arc at the equator on the earth's surface, assuming the radius of the earth to be 3963 miles.
5. Two points $A$ and $B$ are located on a circle of radius 10 ft so that the length of arc $AB$ is 4 ft. Tangents are drawn to the circle at $A$ and $B$. Find the smaller of the angles formed by the intersecting tangents.
6. The sun subtends an angle at the center of the earth of approximately $32'$. The distance from the earth's center to that of the sun is approximately 93,000,000 miles. Find approximately the sun's diameter.
7. An automobile changes its direction of motion by $20^\circ$ while traveling 400 ft around a circular turn in the highway. Find the radius of the circle of the turn.

---

## 3.4 LINEAR AND ANGULAR SPEED OF ROTATION

*Linear speed* is defined as distance traversed in a unit of time. The motion may be along a straight line or along a curve. *Angular speed* is rate of rotation given in angular units per unit of time. If an object moves with a constant linear speed of $v$ during a time interval $t$, the distance is

$$d = vt$$

If an object rotates with a constant angular speed of $\omega$ radians per unit of time, the angle through which it turns in time $t$ is

$$\theta = \omega t$$

In circular motion, $d =$ arc length $s$, so that (3.2) becomes

$$d = \theta r$$

Substituting the above values of $d$ and $\theta$ in this equation gives

$$vt = \omega t r$$

If both members are divided by $t$, this becomes

$$v = \omega r \qquad (3.3)$$

In words: The linear speed of an object moving in a circular path is equal to the magnitude of the angular speed multiplied by the length of the radius of the circular path. This relation is useful in the calculation of outward (centipetal) force tending to make a rapidly rotating object fly apart or leave the circular path.

**EXAMPLE**

A car is traveling with a speed of 30 mph. The tread diameter of its wheels is 27 in. Find the speed of rotation of its wheels in radians per second; in revolutions per second.

**Solution:** It is useful to know that a speed of 60 mph is the same as a speed of 88 ft per sec. (Verify this.) Hence 30 mph is equal to 44 ft per sec. We have $r = \frac{1}{2}(27)/12 = \frac{9}{8}$ ft, and $v = 44$ ft per sec. Also, from (3.3),

$$\begin{aligned}\omega &= \frac{v}{r} \\ &= \frac{44}{\frac{9}{8}} \\ &= 44(\tfrac{8}{9}) \\ &\approx 39.1 \text{ rdn per sec}\end{aligned}$$

Since 1 revolution equals $2\pi$ rdn, we have

$$\begin{aligned}\omega &= \frac{39.1}{2\pi} \text{ rps} \\ &\approx 6.3 \text{ rps}\end{aligned}$$

## *EXERCISES*

1. An automobile traveling at the speed of 4 mph has wheels 30 in. in diameter. Find the angular velocity of the wheels in revolutions per second.
2. Find the angular speed of the minute hand of a clock.
3. Two pulleys are connected by a belt. The small pulley has a diameter of 4 in. If the radius of the larger pulley is 5 in., find the ratio of the angular speed of the small pulley to that of the larger one when the belt is in motion.
4. An object is orbiting the earth with an average speed of 17,000 mph. The path is approximately a circle of radius 4200 miles. Find the angular speed in radians per hour, in degrees per hour.

## 3.5 TRIGONOMETRIC FUNCTIONS OF NUMBERS

Recall that the sine function, sine: $A \to T$ assigned to each angle $\theta$ of the set of all angles $A$ a number ($y/r$, from the standard position setup) called $\sin \theta$ that was an element of the set of numbers $R$. We identified each angle by its measure, since to each angle there is exactly one measure (using a given unit of measure). Thus sin 2 was understood to mean the number assigned to an angle of 2 rdn by the sine function. The domain was the set of all angles, the range, a set of numbers, the interval $[-1, 1]$.

The question arises, "May we regard sin 2 (approximately 0.9086, from Table V) as the number assigned to the *number* 2 instead of to an *angle* of 2 rdn?" More generally, "Why not have a sine function whose domain is the set of (real) numbers?" The answer is that there is no obstacle to doing this and that it is advantageous in many applications of trigonometry. To this end, let us define the function

$$\text{sine}: R \to R$$

as assigning to each number $t$ belonging to the set of numbers $R$ a number $\sin t$, also belonging to $R$, which is given by the equation

$$\sin t = \sin (\text{angle of measure } t \text{ rdn})$$

With this understanding, we may now regard such expressions as sin 2 or $\sin \pi/2$ as being the numbers 0.9086 or 1, respectively, whether 2 and $\pi/2$ represent real numbers or angles in radian measure. The interpretation will depend upon the mathematical situation giving rise to their use.

The above statements about the sine function of numbers apply with equal generality to the cosine, tangent, cotangent, second, and cosecant functions.

We remark that logically, we could have defined $\sin t$ as being equal to sin (angle of measure $t$ deg). However, this gives rise to an ambiguity that would require repeated clarification as to which value of $\sin t$ we are dealing with. The extensive use of radian measure in calculus and its applications causes us to choose the above definition,

$$\sin t = \sin (\text{angle of } t \text{ rdn})$$

with similar definitions for $\cos t$, and so forth.

Finally, we note that we have not caused any confusion between degree measure and radian measure, so that "sin 4°" and "sin 4" are clearly determined and distinct numbers.

Some writers define the trigonometric functions of numbers in terms of number of units of arc length. We have related the number to number of units of radian measure in an angle because of the wide use of the trigonometric functions as functions of angles.

## *MISCELLANEOUS EXERCISES/CHAPTER 3*

**1.** Prove that $\theta \text{ rdn} = \left(\frac{180}{\pi}\theta\right)^\circ$. (This means that to change the radian measure of an angle to degree measure, multiply the radian measure by $180/\pi$.

**2.** Use the formula of Exercise 1 to change the following radian measures to degree measures.

(**a**) $\frac{\pi}{4}$ (**c**) $3\pi$ (**e**) $\frac{4\pi}{3}$

(**b**) $-\frac{2\pi}{3}$ (**d**) $-2\pi$ (**f**) $\frac{7\pi}{4}$

**3.** As in Exercise 1, show that $\theta° = (\pi/180)\theta$ rdn. What does this say about converting from degree to radian measure?

**4.** Use the formula of Exercise 3 to convert the following to radian measure.

(**a**) $72°$ (**c**) $1°$ (**e**) $330°$

(**b**) $-105°$ (**d**) $240°$ (**f**) $500°$

**5.** Use $\pi = 3.14159$ to show that $1° = 0.017453$, correct to five significant digits, and that $1 \text{ rdn} = 57.296° = 57°18'$, correct to figures given.

**6.** Note that Table V may be used to evaluate $\sin x$, regardless of whether $x$ is a number or an angle whose measure is $x$ rdn. However if a table such as Table V is not available, one would proceed to evaluate an expression such as $\sin \pi/12$ as follows:

$$\sin (\text{the number } \pi/12) = \sin (\pi/12 \text{ rdn}) = \sin 15° \approx 0.2588. \text{ (Table III)}$$

Evaluate the expression $\cos kt$, where $k = 0.14$ and $t = 5$, first using Table V and then using Table III. The results should, of course, be approximately equal.

**7.** If the radius of a circle is $r$, the central angle of a sector is $\theta$ rdn (see the figure), and the area of the sector is $K$, prove that

$$K = \frac{1}{2}r^2\theta$$

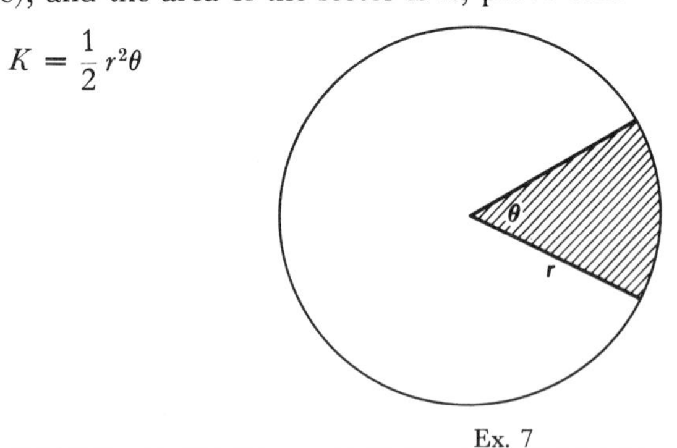


Ex. 7

(*Hint:* From plane geometry, the areas of sectors of a circle are proportional to their central angles. Hence

$$\frac{\text{area of sector}}{\text{angle of sector in radians}} = \frac{\text{area of circle}}{2\pi}\Bigg)$$

**8.** Find the area $K$ of the circular sector determined by:

(**a**) $r = 4$ ft, $\theta = 4$ rdn

(**b**) $r = 3$, $\theta = \dfrac{\pi}{2}$ rdn

(**c**) $r = 10$, $s$ (the length of arc intercepted by the radii) $= 15$

(**d**) $r = a$, $\theta = b°$

**9.** Let $s$ be the arc intercepted on a unit circle by a central angle of $\theta$ radians. Let $t$ be the length of the portion of the tangent line at one extremity of the arc shown in the illustration. If $0 \leqq \theta < \pi/2$, prove that

$$s \leqq t$$

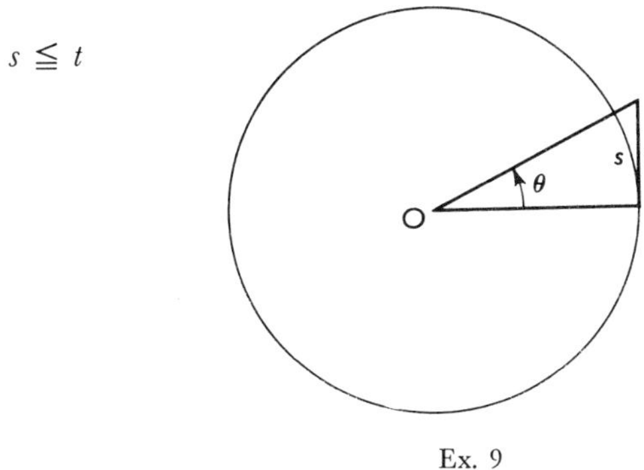


Ex. 9

For what value of $\theta$ does the equality hold? (*Hint:* Compare the area of the sector with the area of the triangle within which the sector lies.)

**10.** Make the assumptions that the earth moves around the sun once in 365 days and that its orbit is circular, with 93,000,000 mile radius. Find an approximation to the linear speed of the earth in its orbit in miles per hour.

**11.** Two points on the earth's surface have latitudes 30° north and 45° north, respectively. Their longitude is the same. Assuming the radius of the earth to be 4000 miles, find the length of arc $AB$.

# CHAPTER 4. VARIATION and GRAPHS of the TRIGONOMETRIC FUNCTIONS

## 4.1 THE SINE AND THE COSINE

In the previous chapters there are a number of exercises giving practice in finding the values of sin $\theta$, cos $\theta$, and so forth, for values of $\theta$ chosen at random. Consider now the manner in which sin $\theta$ varies as $\theta$ runs continuously through a given interval of values. As an aid to visualization, we construct a line segment whose length and direction represents sin $\theta$ for a given $\theta$. If on the unit circle (with center at the origin), the point $P(x, y)$ is located so that $\angle XOP$ is an angle of $\theta$ radians, then

$$\sin \theta = \frac{y}{1} = y$$

If $\overline{MP}$ is the perpendicular segment from $P$ to the $x$ axis, then $|y| = MP$. Furthermore, $y$ is positive or negative according to whether $P$ is above or below the $x$ axis. Hence, when the direction of $P$ from $M$ is considered, we may regard $\overline{MP}$ as representing the number sin $\theta$. The "directed segment" $\overline{MP}$ is called a *line value of sin $\theta$*.

In a similar manner, since

$$\cos \theta = \frac{x}{1} = x$$

we may represent cos $\theta$ by the segment $\overline{0M}$, the associated sign being plus or minus according as $M$ is to the right or to the left of 0.

In each of the parts of Figure 4.1, line values of sin $\theta$ and cos $\theta$ are shown for a single value of $\theta$ in the different quadrants. Imagine watching an animated cartoon in which $\theta$ varies continuously from 0 to $2\pi$, and thus visualize how sin $\theta$, represented by $\overline{MP}$, increases from 0 to 1, at $\theta = \pi/2$. From there sin $\theta$ decreases to 0, at $\theta = \pi$; it decreases to $-1$, at $\theta = 3\pi/2$; thence it increases to 0, at $\theta = 2\pi$. If $\theta$ continues to increase from $2\pi$ to $4\pi$,

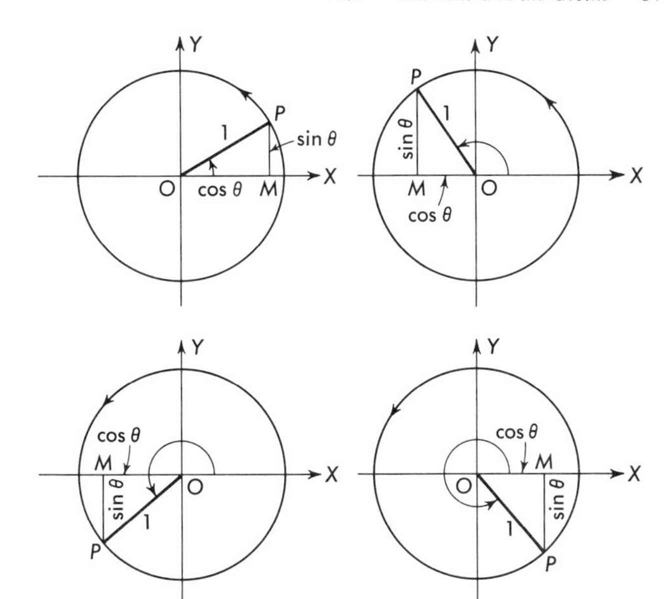


Fig. 4.1 Line values of sin $\theta$ and cos $\theta$

sin $\theta$ repeats the variation of values described above. If $\theta$ continues to increase indefinitely, sin $\theta$ periodically repeats the same set of values for every addition of $2\pi$ to the interval of $\theta$. There is corresponding periodic variation for negative values of $\theta$.

If $y = \sin \theta$, the total set of points $(\theta, y)$, plotted on a rectangular coordinate system, constitutes the graph of the sine function. The result for $-5.5 < \theta < 6.8$ appears in Figure 4.2.

A similar study of the variation of cos $\theta$ may be made by observing in Figure 4.1 the changing values of $\overline{OM}$, representing cos $\theta$ as $\theta$ varies continuously. The graph of $y = \cos \theta$, also shown in Figure 4.2, is seen to differ

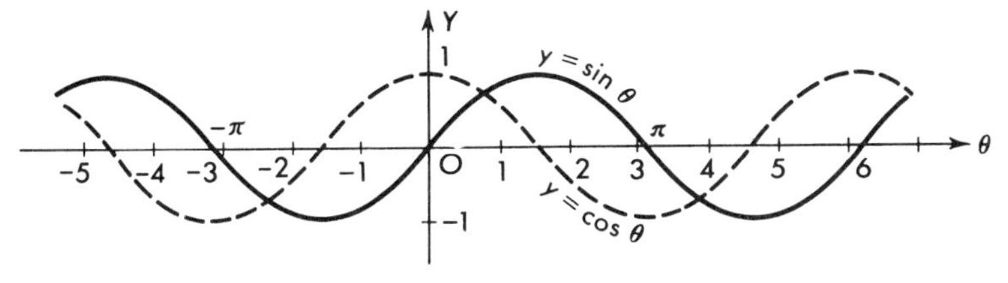


Fig. 4.2 Graphs of $y = \sin \theta$ and $y = \cos \theta$

from that of $y = \sin \theta$ only in its position along the $\theta$ axis. A sliding (or translation) of the cosine curve through a distance of $\pi/2$ units along the $\theta$ axis makes it coincide with the sine curve. This is algebraically stated by the relation

$$\sin\left(\theta + \frac{\pi}{2}\right) = \cos \theta$$

an equation that we shall prove later (Exercise (5b), Section 6.9) is satisfied by every value of $\theta$.

The graph of $y = \sin \theta$ may be traced quickly and with reasonable accuracy by using line values of $\sin \theta$. Construct a unit circle with its center at any convenient point of the $\theta$ axis (see Figure 4.3). Choose a value of $\theta$,

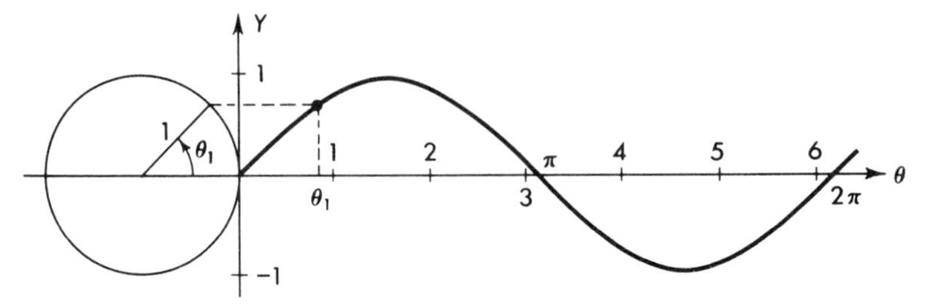


Fig. 4.3 Graph of $y = \sin \theta$ from line values

say $\theta_1$, for which the corresponding point on the graph is desired. Construct an angle of measure $\theta_1$ rdn as a central angle in the circle and locate on the $\theta$ axis the point whose coordinate is $\theta_1$. For instance, if $\theta_1 = 2$, locate the point with abscissa 2 on the $\theta$ axis. Through the point at which the terminal side of $\angle\theta_1$ intersects the circle, draw a horizontal line, and through the point $\theta = \theta_1$ on the $\theta$ axis draw a vertical line. The point of intersection of these two lines is the point given $(\theta_1, \sin \theta_1)$, since its abscissa is $\theta_1$ and its ordinate is given by the line value of $\sin \theta_1$. This can be repeated for as many values of $\theta$ as are needed to trace the curve. The figure shows the resulting curve for $0 \leqq \theta \leqq 2\pi$, approximately.

If the unit circle is rotated counterclockwise through 90°, as in Figure 4.4, the ordinate of the point located by the intersection of the corresponding horizontal and vertical lines is the line value of $\cos \theta_1$. Hence the locus of points so obtained for all values of $\theta$ is the graph of $y = \cos \theta$. This rotation of the circle through 90°, or $\pi/2$ rdn, again calls attention to the relation $\sin (\pi/2 + \theta) = \cos \theta$.

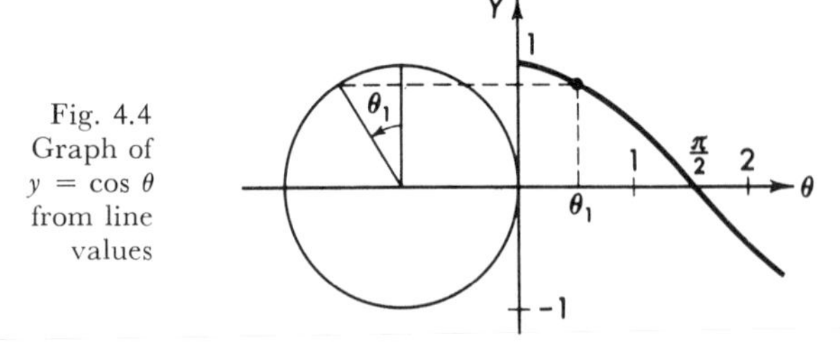


Fig. 4.4 Graph of $y = \cos \theta$ from line values

A function that repeats its values in the fashion of $\sin \theta$ or $\cos \theta$ is called a *periodic* function. More precisely, a function $f(x)$ is *periodic of period* $p$ if $p$ is the smallest positive number for which $f(x + p) = f(x)$ for all values of $x$, assuming that such a number $p$ exists. We have seen that $\sin (\theta + 2\pi) = \sin \theta$ and that $\cos (\theta + 2\pi) = \cos \theta$, for all values of $\theta$. Accepting the fact that $2\pi$ is the smallest number for which these relations hold, it can be said that $\sin \theta$ and $\cos \theta$ are periodic functions, both of period $2\pi$. By the *amplitude* of a periodic function $y = f(x)$ is meant the maximum value of $|f(x)|$, if this maximum value exists. This is the greatest deviation of any point of the curve from the axis. The amplitude of $y = \sin \theta$ is 1, as is that of $y = \cos \theta$.

From the standpoint of the graph of a periodic function, a *cycle* is the shortest possible segment of the curve such that the entire curve is a repetition of this segment. Figure 4.3 shows a little more than one cycle of $y = \sin \theta$.

The wavelike sine and cosine curves play an important role in applied mathematics, particularly in the study of wave motions and periodic phenomena such as alternating currents and vibrating strings. For example, the graph of an alternating current as a function of the time* is simply a sine curve, usually with 60 cycles to 1 sec of time. Since there are an immense number of cycles in a few hours, we see the need to appreciate the seemingly endless repetition of the curve $y = \sin \theta$. The scale of Figure 4.5 is small enough to emphasize this.

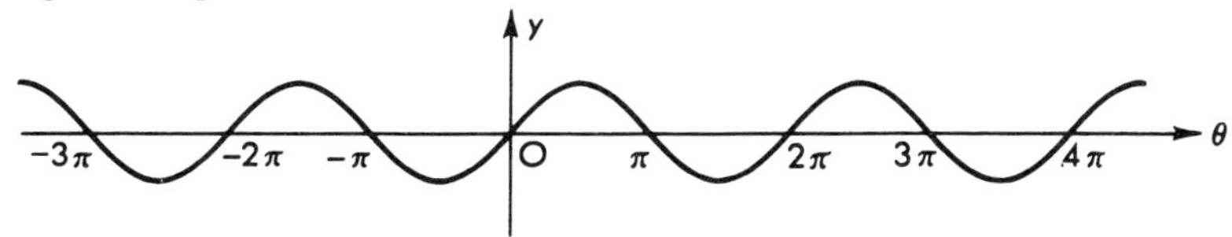


Fig. 4.5 $y = \sin \theta$, showing several cycles

Having viewed the graph of $y = \sin \theta$ "in the large," with the idea of noting the over-all variation, let us now have a close look at the first quarter of the cycle, starting with $\theta = 0$. Again note how the curve rises from the point $(0, 0)$ to the point $(\pi/2, 1)$. The precise way in which the curve rises over this interval is important in obtaining a true picture of this portion of the cycle. A careful plotting of the curve over the interval $0 \leqq \theta \leqq \pi/2$ reveals the correctness of the sketch in Figure 4.6 in showing this quarter-cycle of

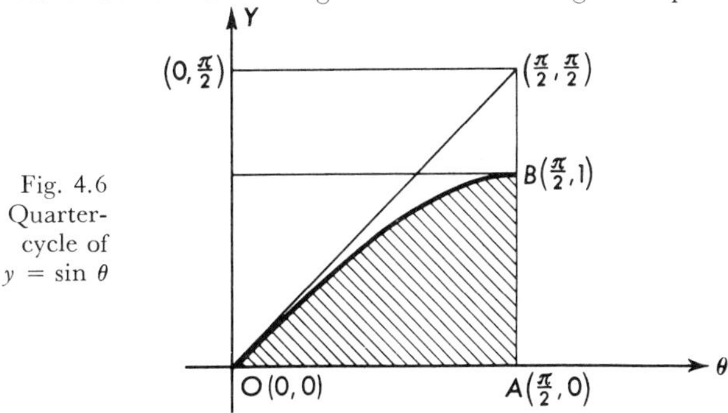


Fig. 4.6 Quarter-cycle of $y = \sin \theta$

$y = \sin \theta$ as related to a square of sides $\pi/2$ (approximately 1.57) units long. Note that at $(0, 0)$ the curve is tangent to the diagonal of the square (you will

* Illustrating one situation in which $\theta$ is a number, not an angle.

be asked to verify this in Section 4.6), and that at the point $(\pi/2, 1)$ it is tangent to the horizontal line through that point.

A cutout of the shaded portion of Figure 4.6 serves as a pattern from which the entire curve for $y = \sin \theta$, or that for $y = \cos \theta$, may be traced. Turn the cutout over, revolving about the line $\overleftrightarrow{AB}$, and it will be in position to trace the second quarter-cycle. The manner in which it must be placed to continue tracing the curve indefinitely in either direction becomes evident by referring to Figure 4.3 or 4.5. This pattern also fits the quarter-cycle of $y = \cos \theta$ shown in Figure 4.4, if the scales agree, and the entire curve can be traced with this pattern.

## *EXERCISES*

**1.** From Table V, complete the table of sine values below.

| $\theta$ | 0 | 0.2 | 0.4 | 0.6 | 0.8 | 1.0 | 1.2 | 1.4 | 1.6 |
|---|---|---|---|---|---|---|---|---|---|
| $\sin \theta$ | | | | | | | | | |

On a coordinate system plot the nine points whose coordinates are the ordered pairs $(\theta, \sin \theta)$ given in your table. Connect these points consecutively with a smooth curve. Compare this with Figure 4.6, noting that $\pi/2 \approx 1.6$.

**2.** Without tables show that $\sin \pi/6 = \sin 5\pi/6$, $\sin \pi/4 = \sin 3\pi/4$, $\sin \pi/3 = \sin 2\pi/3$, and $\sin 0 = \sin \pi$. Does this help you to see that the pattern in which the sine curve rises on the interval $[0, \pi/2]$ is exactly reversed as the curve descends on the interval $[\pi/2, \pi]$?

**3.** As in Exercise 2, show that $\sin 7\pi/6 = -\sin \pi/6$, $\sin 5\pi/4 = -\sin \pi/4$ $\sin 5\pi/3 = -\sin \pi/3$, and $\sin 3\pi/2 = -\sin \pi/2$. This repetition of values with opposite signs should indicate the "cutout" suggested from Figure 4.6 can be used in tracing the graph of the sine function as $\theta$ runs through the third quadrant.

**4.** Cut out the curve you sketched in Exercise 1, as suggested, and try tracing the sine curve through a couple of cycles.

**5.** A mathematics project seen at science fairs is a sine curve drawing machine. Turn a crank and it traces out a sine curve. Can you design one?

## 4.2 THE GRAPHS OF $y = A \sin Bx$ AND $y = A \cos Bx$*

The ordinate ($y$ value) of each point of the graph of $y = A \sin x$, where $A$ is a constant, is $A$ times the corresponding ordinate of $y = \sin x$. The graph of

* The choice of $\theta$ as a symbol for the variable in Section 4.1 was based on the variation of an angle. The fact that the variable may be a *number* and the fact that the horizontal axis is ordinarily called the $X$ axis now lead to symbolizing the variable with the letter $x$.

$y = A \sin x$ is thus the graph of $y = \sin x$ *stretched or shrunk in the Y direction by the multiple A*. One cycle of the curve for $y = 3 \sin x$ is shown in Figure 4.7, which includes the graph of $y = \sin x$ for comparison. The period of both functions is the same, namely, $2\pi$.

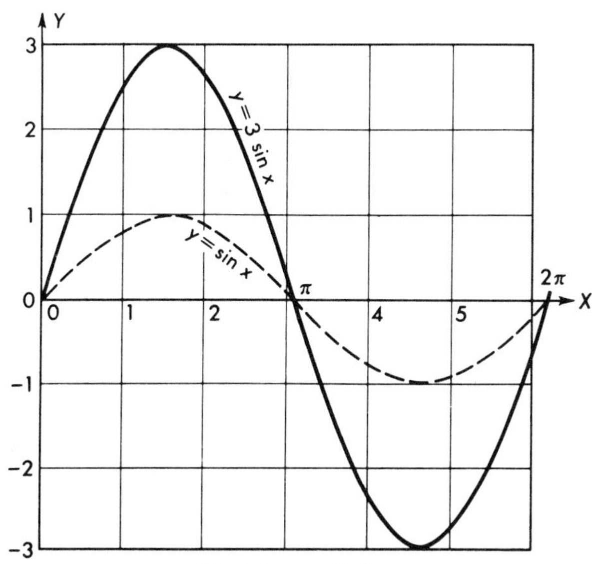


Fig. 4.7 $y = 3 \sin x$

The graph of $y = A \cos x$ is also the graph of $y = \cos x$ stretched in the $Y$ direction by the multiple $A$.

A complete cycle of the curve for $y = 3 \sin 2x$ is completed as $2x$ increases from 0 to $2\pi$. Hence the period for this function is the solution of the equation

$$2x = 2\pi$$

that is,

$$x = \pi$$

Thus the graph of $y = 3 \sin 2x$ is the graph of $y = 3 \sin x$, squeezed horizontally by the multiple $\frac{1}{2}$. The graph of $y = 3 \sin 2x$ is shown in Figure 4.8. It should be compared with Figure 4.7.

In general, the period of both $A \sin Bx$ and $A \cos Bx$ is $2\pi/|B|$. The amplitude of each is $|A|$. A knowledge of the period and amplitude of these functions, combined with the fact that the entire curve is composed of rearrangements of the first quarter-cycle, starting with $x = 0$ (as pointed out in connection with the "cutout" scheme of sketching $y = \sin \theta$), permits the making of a rapid sketch by the following steps:

(1) Measure off one period, $\dfrac{2\pi}{|B|}$, on the $X$ axis beginning at the origin.

(2) Divide this period into four quarter-periods.

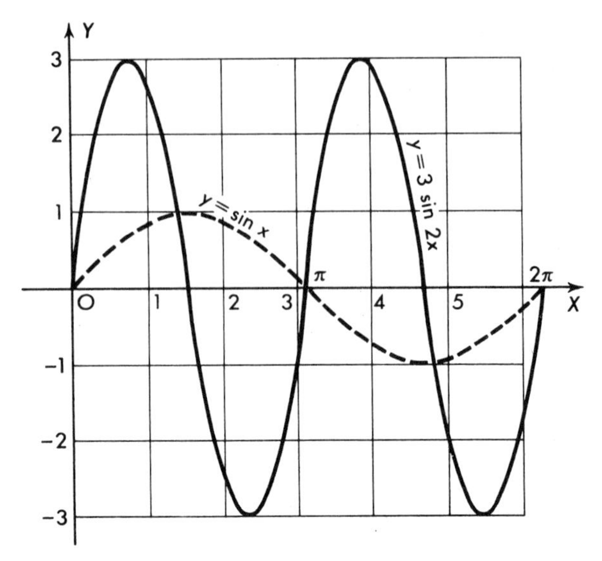


Fig. 4.8 $y = 3 \sin 2x$

(3) On the $Y$ axis, locate the points where $y = A$ and $y = -A$, and through them draw light lines parallel to the $X$ axis. Block off the period and quarter-periods with light lines parallel to the $Y$ axis.

(4) Remembering that $\sin 0 = 0$ and that $\cos 0 = 1$, start the sketch of $y = A \sin Bx$ at the origin, or the sketch of $y = A \cos Bx$ at the point $(0, A)$. Fill in the first quarter-cycle, perhaps thinking of a "stretched" or "squeezed" version of the curve in Figure 4.6.

(5) Complete the cycle by following through with the pattern established in the first quarter-cycle.

(6) Repeat for as many cycles as desired.

**EXAMPLE**

Sketch the graph of $y = -4 \cos \pi x$ through several cycles.

**Solution:** The period is $2\pi/\pi = 2$. The amplitude is 4. Divide the period from $x = 0$ to $x = 2$ into four quarter-periods, constructing perpendiculars to the $X$ axis at the division points, as shown in Figure 4.9. Draw bounding lines parallel to the $X$ axis 4 units (the amplitude) above and below it. Note that, when $x = 0$, $y = -4 \cos 0 = -4$. Plot the point $(0, -4)$. Remembering the manner of variation of the cosine function, sketch the first cycle in the blocked space. (If necessary, check with a few values of $x$, say $x = \frac{1}{2}$, for which $y = -4 \cos \frac{1}{2}\pi = 0$. The point $(\frac{1}{2}, 0)$ is on the curve.) Using the first cycle as a pattern, sketch as many additional cycles in each direction as desired. In Figure 4.9 the first cycle is drawn heavily to call it to the attention of the reader. Normally, the entire curve is represented uniformly.

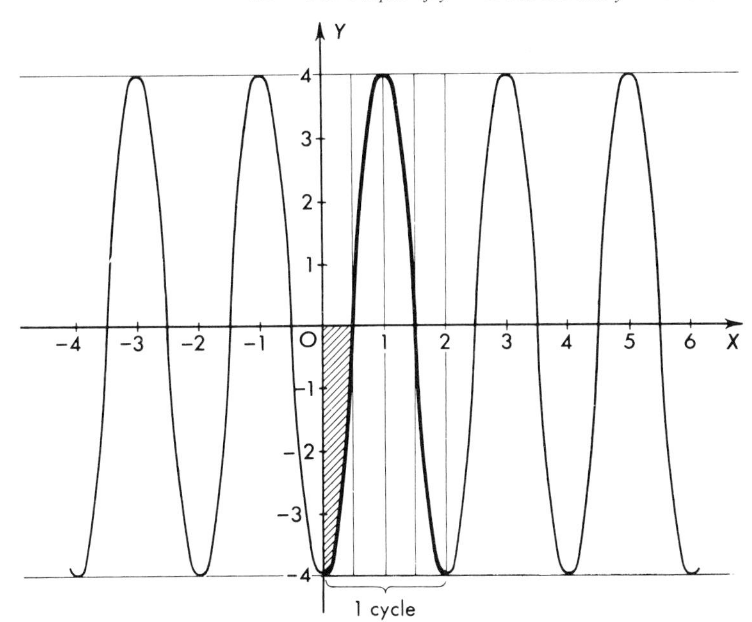


Fig. 4.9 Rapid sketch of $y = -4 \cos \pi x$

## *EXERCISES*

1. A zero of a function is a value of its variable for which the value of the function itself is zero. For example, a zero of the function $\sin \theta$ is $\pi$, because $\sin \pi = 0$. Actually, $\sin \theta$ has an endless number of zeros. They are: $0, \pm\pi, \pm 2\pi, \pm 3\pi, \ldots$, or, in general, $n\pi$, where $n = 0, \pm 1, \pm 2, \pm 3, \ldots$ Referring to the graph of $\cos \theta$ in Figure 4.2, give a general expression for the zeros of $\cos \theta$. Repeat for the function $-4 \cos \pi x$, pictured in Figure 4.9.
2. Referring again to Figure 4.2, state what values of $\theta$ make $\sin \theta = \cos \theta$. What are the zeros of the function $\sin \theta - \cos \theta$?
3. Sketch the graph of each of the following functions through at least two cycles, starting at a point of your choice. Give the period and amplitude of each.

   (**a**) $y = 2 \cos x$

   (**b**) $y = 2 \sin 3x$

   (**c**) $y = 3 \cos \dfrac{\pi x}{2}$

   (**d**) $y = -3 \sin 2x$

   (**e**) $y = \frac{1}{2} \sin \dfrac{x}{2}$

   (**f**) $y = \frac{3}{2} \cos \pi x$

4. Sketch each of the following functions on the specified $x$ interval.

   (**a**) $y = 2 \cos 4x,\ 0 \leqq x \leqq \dfrac{3\pi}{2}$

   (**b**) $y = 3 \sin \pi x,\ -2 \leqq x \leqq 2$

   (**c**) $y = \frac{8}{3} \sin 3x,\ 0 \leqq x \leqq \pi$

   (**d**) $y = 2 \cos \dfrac{x}{2},\ -2\pi \leqq x \leqq 2\pi$

5. Functions of the type $y = \sin(x + \pi/2)$ can be sketched without laborious point plotting if one notices that to obtain a certain value of $y$, $x$ is chosen $\pi/2$ less than the value of $x$ necessary to give the same value of $y$ from the now familiar function $y = \sin x$. For instance $x = -\pi/2$ corresponds to $y = 0$ from $y = \sin(x + \pi/2)$, while $x = 0$ corresponds to $y = 0$ from $y = \sin x$. Since this "shift" applies for all points, it is only necessary to sketch $y = \sin x$, translated (shifted) to the left by $\pi/2$ units to obtain the sketch of $y = \sin(x + \pi/2)$. Use this suggestion to sketch each of the following through one cycle.

   **(a)** $y = \sin(x - \pi/2)$. Think which direction you must translate
   **(b)** $y = \cos(x + 2)$
   **(c)** $y = 2 \sin 2(x + \pi/2)$
   **(d)** $y = 3 \cos(\pi x - \pi)$ (Rewrite in the form of part **(c)**)

6. In applied mathematics, periodic phenomena are studied. The trigonometric functions are of importance in describing the behavior of repeating physical quantities. Alternating currents (usually 60 cycles per second), vibratory motion of beams or strings, pulsating forces, and so forth come under this head. Constants involved may be large relative to those we have used in the above exercises. A suitable choice of scales on the coordinate axes will permit sketches of convenient size. Sketch the functions below, adjusting scales as you think advisable.

   **(a)** $y = 25 \sin 40\pi t$
   **(b)** $y = 100 \cos 200t$

---

## 4.3 THE TANGENT AND COTANGENT

In considering the variation of the function $\tan\theta$, it is instructive to obtain line values for $\tan\theta$, at least for $0 \leqq \theta \leqq \pi/2$. Let the angle of $\theta$ radians be in standard position (Fig. 4.10). At the point $A(1, 0)$ erect line $AB$ perpendicular to the $X$ axis, the terminal side of $\theta$ intersecting $AB$ at the point $P(1, y)$. Then

$$\begin{aligned}\tan\theta &= \frac{y}{1}\\ &= y\\ &= AP \text{ throughout quadrant I}\end{aligned}$$

As a check, we note that, for $\theta = 0$, $AP = 0$, in agreement with our knowledge that $\tan 0 = 0$. At $\theta = \pi/4$, $AP = 1$, which checks with $\tan\pi/4 = 1$. Note that, as $\theta$ increases toward $\pi/2$, $\tan\theta = AP$ increases without bound. At $\theta = \pi/2$, the terminal side of $\theta$ is parallel to $AB$, so that no point $P$ is determined. Thus, again we say that $\tan\pi/2$ (or $\tan 90°$) is meaningless, or that it is undefined. (*Note:* If a unit circle is superposed on Figure 4.10 with center at $O$, then $AP$ is tangent to the circle; hence the name of the trigonometric function tangent of $\theta$.)

Figure 4.11 displays the graph of $y = \tan x$. Tan $x$ increases as $x$ increases on the interval $-\pi/2 < x < \pi/2$. The continuous (unbroken) curve cor-

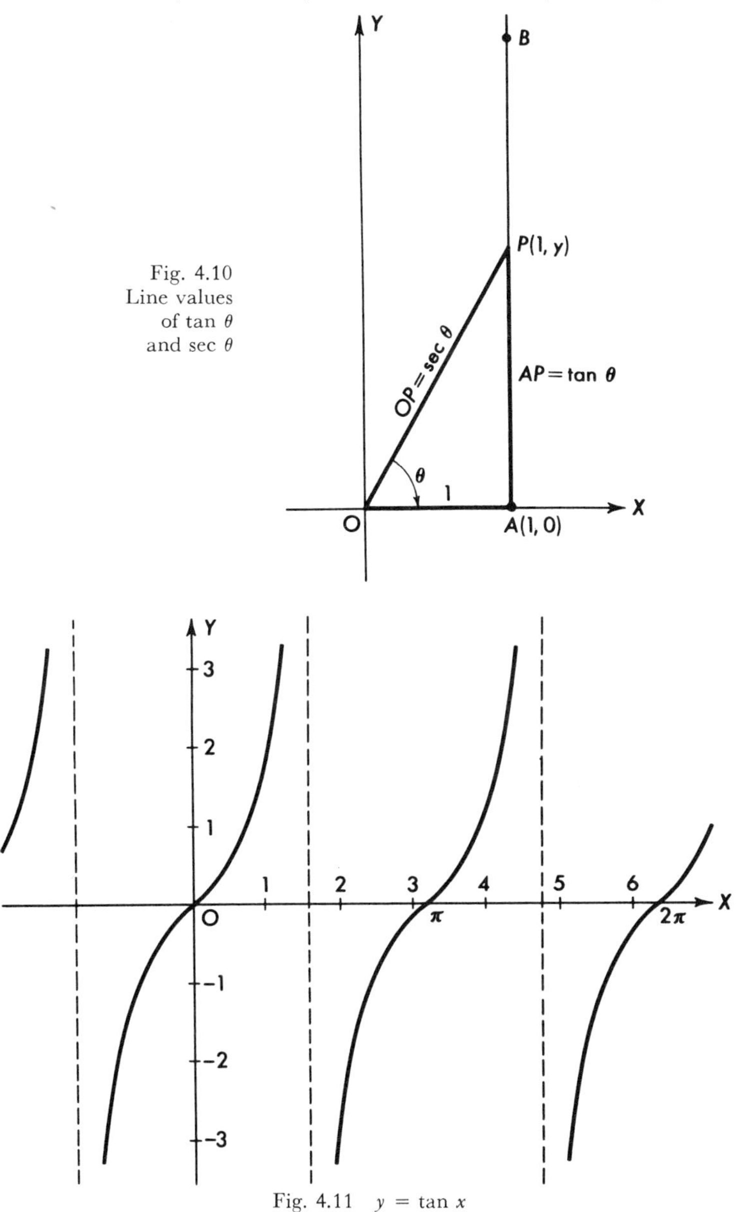


Fig. 4.10 Line values of tan $\theta$ and sec $\theta$

Fig. 4.11 $y = \tan x$

responding to this interval proves to be a cycle of the entire curve, since the remaining portions are repetitions of this part. Tan $x$ is then periodic of period $\pi$, in contrast to the period of $2\pi$ for sin $x$ and for cos $x$. Tan $x$ is said to be discontinuous (its graph is broken) at $x = \pi/2$, $x = 3\pi/2$, and, in general, at $x = n\pi/2$, where $n$ is any odd integer.

The graph of $y = \cot x$ (Figure 4.12) is similar to that of $y = \tan x$. As $x$ increases over the interval $0 < x < \pi$, cot $\theta$ *decreases*. Cot 0 is meaningless,

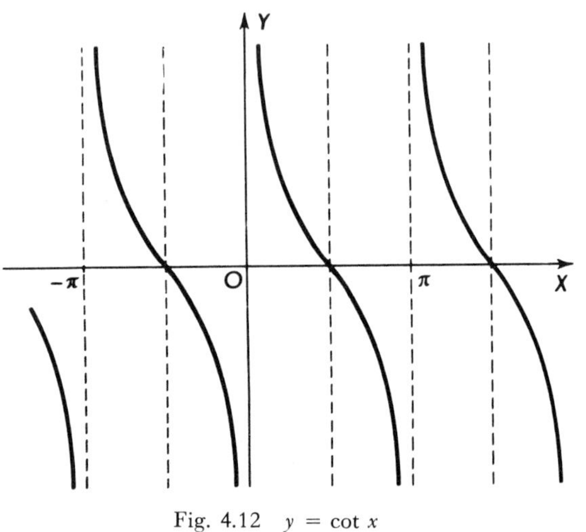


Fig. 4.12 $y = \cot x$

as is cot $\pi$. Why? The period of cot $x$ is $\pi$. An amplitude for tan $x$ or for cot $x$ cannot be specified as there is no maximum absolute value of either.

## 4.4 THE SECANT AND THE COSECANT

In Figure 4.10, the definition of sec $\theta$ gives in quadrant I.

$$\sec\theta = \frac{OP}{1} = OP$$

so that $OP$ is a line value of sec $\theta$ for $O \leqq \theta < \pi/2$. As $\theta$ increases from zero, sec $\theta$ increases, and, as $\theta$ nears $\pi/2$, sec $\theta$ increases without bound. Figure 4.13 shows a sketch of the graph of $y = \sec x$. Sec $\pi/2$ is meaningless, as is sec $n\pi/2$, where $n$ is any odd integer. The graph of $y = \cos x$ is sketched on the same set of axes, in order to display the relation

$$\sec x = \frac{1}{\cos x}$$

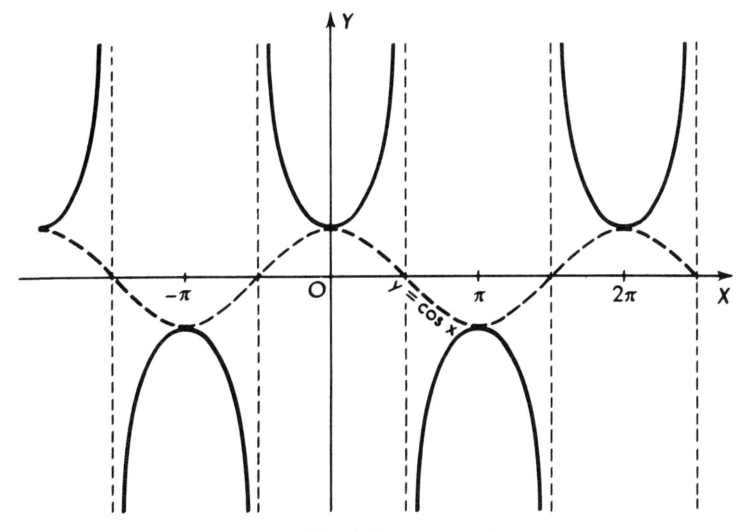


Fig. 4.13 $y = \sec \theta$

which will be verified in Chapter 6. For example, at $x = \pi/3$, $\cos x = \frac{1}{2}$, so that $\sec x = 1/\frac{1}{2} = 2$. Thus the ordinate of every point of the curve $y = \sec x$ may be determined as the reciprocal of the ordinate of the corresponding point on the $y = \cos x$ curve except for the points at which $\cos x = 0$, where $\sec x$ is meaningless. Note that $|\sec x| \geqq 1$ for every value of $x$ for which $\sec x$ is defined, while $|\cos x| \leqq 1$ for every value of $x$. The period of $\sec x$ is $2\pi$.

The graph of $y = \csc \theta$ bears the same relation to the graph of $y = \sin \theta$ that the graph of $y = \sec \theta$ does to the graph of $y = \cos \theta$. If the curves of Figure 4.13 are translated $\pi/2$ units horizontally to the right, the figure becomes the graph of $y = \csc \theta$, together with that of $y = \sin \theta$. It is left to the student (see the exercises below) to sketch the graph of $y = \csc \theta$.

## *EXERCISES*

**1.** State what values of $x$, if any, are zeros of each of the following functions:

(**a**) $\tan x$ (**c**) $\sec x$ (**e**) $2x + 7$
(**b**) $\cot x$ (**d**) $\csc x$ (**f**) $x + \sin x$

**2.** Sketch the graph of $y = \csc x$ for the interval $-\pi < x < 2\pi$.
**3.** Sketch the graph of $y = -\tan x$ for $0 \leqq x \leqq 2\pi$.
**4.** Sketch the graph of $y = -\sec x$ for $-\pi/2 < x < 3\pi/2$.
**5.** Sketch the graph of $y = \tan 2x$ for $0 \leqq x \leqq \pi$.

**6.** Sketch the graph through one cycle of

(**a**) $y = \tan (x + 90°)$ (**c**) $y = 3 \tan (2x - 90°)$
(**b**) $y = 2 \cot (x - 90°)$ (**d**) $y = 4 \cot (3x + 90°)$

## 4.5 COMPOSITE FUNCTIONS. ADDITION OF ORDINATES

It is often necessary to consider functions which are composed of sums of simpler functions. Examples are:

$$y = \frac{x}{3} + \sin x \tag{1}$$

$$y = 2 \sin x + \sin 2x \tag{2}$$

$$y = \sin x - \cos x \tag{3}$$

The graphs of such functions can be obtained in the usual manner by substituting values of $x$ to obtain the values of $y$ and plotting the corresponding points. However, considerable labor can be saved if the graphs of the separate functions are first traced, and then are combined into the composite graph by the method of addition of ordinates. The method is best shown by an example. Consider equation (1) above. Trace the loci of the two functions, $y_1 = x/3$ and $y_2 = \sin x$, upon a single set of axes. The first is a straight line and its graph is familiar to the student of algebra.

Since the original function, $y = x/3 + \sin x$, is the sum of two functions, the ordinate of its graph corresponding to a chosen value of $x$ is obviously the *algebraic* sum of the ordinates of the graphs of the separate functions corresponding to the same value of $x$. (The word *algebraic* is used because due regard must be given to the sign of the separate ordinates.) Thus, for $x = 0$, both $x/3$ and $\sin x$ are 0, hence (0, 0) is a point on the composite curve. In general, when $x = a$, $y_1 = a/3$ (ordinate of $B$) and $y_2 = \sin a$ (ordinate of $C$), as in Figure 4.14. Hence the ordinate of the corresponding point $D$ on the

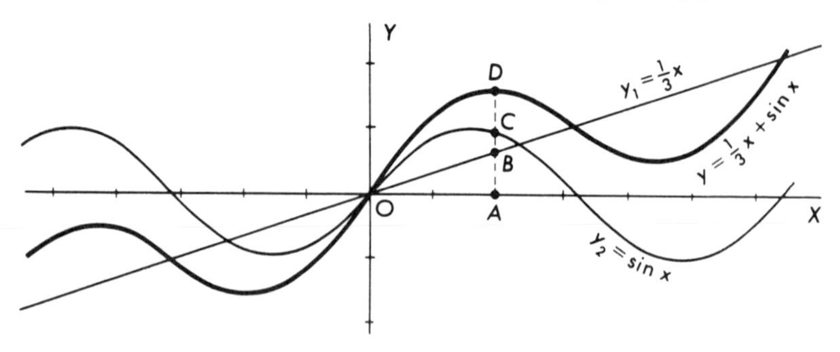


Fig. 4.14 $y = \frac{1}{3}x + \sin x$

composite curve is

$$y = y_1 + y_2 = \frac{a}{3} + \sin a$$

These ordinates can be added graphically, without computing them, by means of dividers, or by marking the edge of a strip of paper to record the lengths. This is repeated to obtain as many points as needed.

The graph of equation (2) $y = 2 \sin x + \sin 2x$, is shown in Figure 4.15 for $0 \leqq x \leqq 2\pi$.

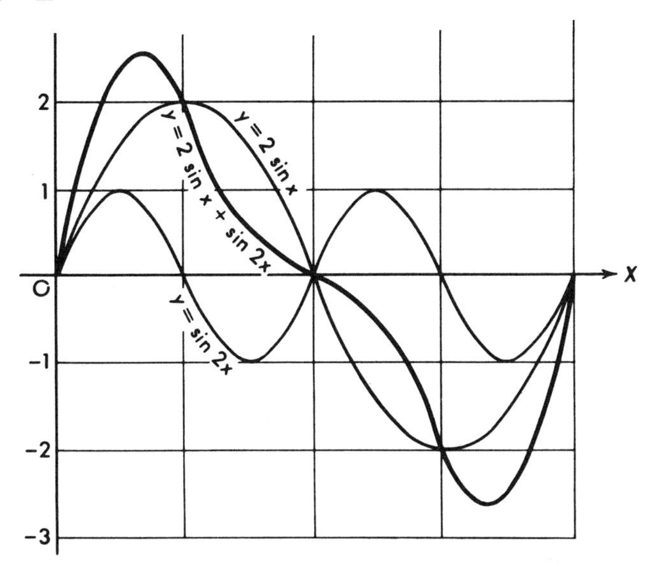


Fig. 4.15 $y = 2 \sin x + \sin 2x$

The graphs of $y = 2 \sin x$ and $y = \sin 2x$ are constructed first, and then the ordinates of the two graphs for the same value of $x$ are added to obtain the corresponding ordinate for the composite graph. It should be noted that the sum of the two trigonometric functions is, in this case, a periodic function whose period is $2\pi$ and consequently the graph may be continued indefinitely to the left and to the right.

## 4.6 THE FUNCTION $\frac{\sin x}{x}$

This function is defined for every value of $x$ except $x = 0$. The question of what happens to the value of $(\sin x)/x$ as $x$ approaches 0, without attaining it, is of great importance in the calculus of trigonometric functions as well as in sketching the graph of $y = (\sin x)/x$.

To study the behavior of $(\sin x)/x$ for small values of $x$, consider Figure 4.16, which shows a positive, acute angle of $\theta$ radians with its vertex at the

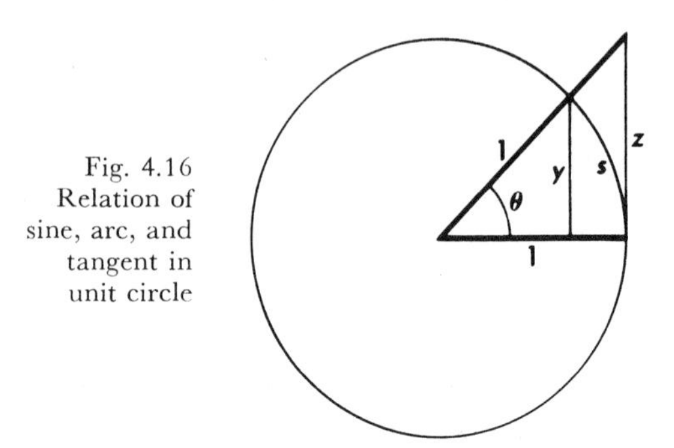


Fig. 4.16 Relation of sine, arc, and tangent in unit circle

center of a unit circle. Then the line values of $\sin \theta$ and $\tan \theta$ are $y$ and $z$, respectively, as shown in the figure. Then (see Exercise 9, Miscellaneous Exercise, Chapter 3),

$$y < s < z$$

But $y = \sin \theta$, $s = \theta$,* and $z = \tan \theta$, so that the inequality becomes

$$\sin \theta < \theta < \tan \theta$$

This inequality is valid if $0 < \theta < \pi/2$, but not if $-\pi/2 < \theta < 0$. Divide the members of this inequality by the positive number $\sin \theta$ and note that $(\tan \theta)/(\sin \theta) = 1/(\cos \theta)$.† We now have the inequality

$$1 < \frac{\theta}{\sin \theta} < \frac{1}{\cos \theta}$$

As $\theta$ nears 0 the value of the fraction $1/(\cos \theta)$ approaches 1. Why? Thus $\theta/(\sin \theta)$ is "squeezed" between 1 and a fraction that approaches 1. *Hence $\theta/(\sin \theta)$ itself approaches 1 as $\theta$ approaches 0*, in the stated range. A similar argument for values of $\theta$ in the range $-\pi/2 < \theta < 0$ (in which case $y$, $s$, and $z$ of Figure 4.16 become negative) results in the same conclusion, for the initial inequality becomes $y > s > z$, and dividing an inequality by the *negative* number $\sin \theta$ *reverses* the inequality. Hence we can say that, as $\theta$ approaches 0, $\theta/\sin \theta$ approaches 1, or that $\sin \theta/\theta$ approaches 1.‡

---

* This is due to the formula of Section 3.2, $s = r\theta$, and the choice of $r = 1$.

† By definition $\tan \theta = y/x$, $\sin \theta = y/r$, and $\cos \theta = x/r$

Hence $$\frac{\tan \theta}{\sin \theta} = \frac{y/x}{y/r} = \frac{r}{x} = \frac{1}{x/r} = \frac{1}{\cos \theta}$$

More of this in Chapter 6.

‡ This informal discussion is intended to be descriptive, not rigorous. In calculus, a more careful treatment using the theory of limits is given.

One way of restating these conclusions is to say that, for small values of $x$, $x/(\sin x) \approx 1$, or that $(\sin x)/x \approx 1$. Another way of putting this is to say that, for small values of $x$, $\sin x \approx x$. For example, $\sin 0.1 = 0.0998$, and $\sin 0.01 = 0.0100000$. In many applied problems $\sin x$ is actually replaced by $x$ in the mathematical treatment, and results valid for small values of $x$ are obtained. For instance, the pendulum formula $t = 2\pi \sqrt{l/g}$, in which $l$, $g$, and $t$ are respectively the length of a pendulum, the acceleration of gravity, and the time of a complete swing of the pendulum, is obtained by replacing $\sin x$ by $x$ at a certain stage of the development, and so is valid only for small swings of the pendulum.

The graph of $y = (\sin x)/x$ is shown in Figure 4.17. By writing the equation in the form $y = (1/x) \sin x$, it is seen that the factor $1/x$ may be

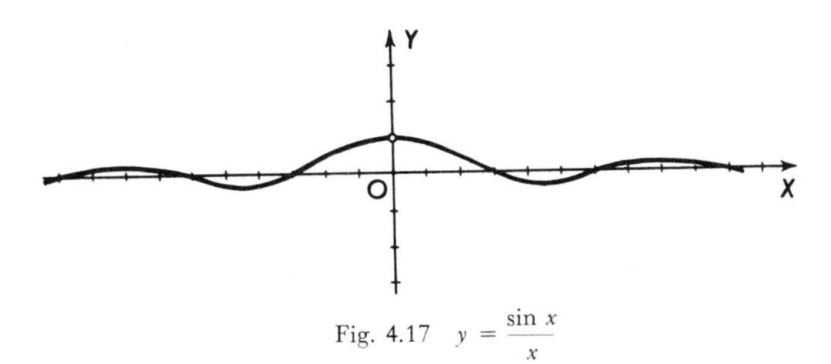


Fig. 4.17 $y = \frac{\sin x}{x}$

regarded as a sort of "variable amplitude" of the function $y = A \sin x$, in which the factor $1/x$ acts as a damping factor to decrease the magnitude of the sine oscillations as the numerical value of $x$ increases. Note that the curve has a missing point at $y = 0$, since $(\sin x)/x$ is not defined there. If it is desired to include this point on the curve, it is necessary to define the function as

$$y = \begin{cases} \dfrac{\sin x}{x} & \text{if } x \neq 0 \\ 1 & \text{if } x = 0 \end{cases}$$

## *EXERCISES*

1. In addition to the six trigonometric functions that are the basic concepts of this book, there are other trigonometric functions that are used occasionally. Three of these are:

$$\text{versed sine } \theta = 1 - \cos \theta, \text{ written vers } \theta$$

$$\text{coversed sine } \theta = 1 - \sin \theta, \text{ written covers } \theta$$

$$\text{haversine } \theta = \frac{1 - \cos \theta}{2}, \text{ written havers } \theta$$

Use the method of addition of ordinates to sketch the graphs of

(**a**) $y = \text{covers } x$
(**b**) $y = \text{havers } x$, each on the interval $-\pi \leqq x \leqq 2\pi$

2. By addition of ordinates, trace on the interval $-\pi \leqq x \leqq 2\pi$ the curves represented by the following equations.

(**a**) $y = \sin x + \cos x$
(**b**) $y = \sin x - \cos x$
(**c**) $y = x + \cos x$
(**d**) $y = 2 \sin x + \sin 2x$
(**e**) $y = \dfrac{x^2}{4} - \sin x$
(**f**) $y = 2 \cos 3x + \sin 2x$

3. Sketch the graph of $y = x \sin x$ on the interval $-2\pi \leqq x \leqq 2\pi$. (*Suggestion:* Sketch the graphs of $y = \sin x$ and the lines $y = x$ and $y = -x$ on one set of axes. Use a method of *multiplication of ordinates* to sketch $y = x \sin x$, noting that, when either factor is zero, $y$ is also zero. Note also that the lines $y = x$ and $y = -x$ form a boundary for the resulting curve, that the curve is tangent to $y = x$ when $\sin x = 1$, and that it is tangent to $y = -x$ when $\sin x = -1$. Thus the factor $x$ in $y = x \sin x$ forms a sort of variable amplitude for the oscillations of $y = \sin x$).
4. Sketch $y = \cos^2 x$ on the interval $0 \leqq x \leqq 2\pi$. Sketch $y = (1 + \cos 2x)/2$ on the same interval. Are these curves actually identical, or just similar in appearance? See Exercise 11, Section 6.11 for an answer.
5. Sketch $y = |\sin x|$ on the interval $0 \leqq x \leqq 2\pi$. Note any important similarity or difference between this curve and the curves of Exercise 4.
6. Find the largest value of $\theta$ in Table V such that $\sin \theta = \theta$ correct to two decimal places. If this maximum value of $\theta$ is the radian measure of an angle, what is the degree measure of $\theta$?
7. Suggest how one can be sure that curve $y = \sin x$ is tangent to the line $y = x$ at the origin. (See Figure 4.6 and discussion.)

---

## 4.7 POINTS IN POLAR COORDINATES

As an angle $\theta$ in standard position increases from 0 to $2\pi$, its terminal side sweeps over the entire plane so that the location of any given point in the plane may be precisely expressed by giving: (**a**) the value of $\theta$ for which the terminal side passes through the point, and (**b**) the signed number $\rho$ giving its distance and direction from the origin. The quantities $\rho$ and $\theta$ which thus locate a point $P$ are called the *polar coordinates* of $P$. The ray in the $\theta = 0$

direction is called the *polar axis;* the origin is the *pole.* In the same way that $(x, y)$ designates a point in rectangular coordinates, $(\rho, \theta)$ represents the point in polar coordinates. The points $(3, \pi/6)$ and $(5, 7\pi/6)$ are shown in the sketches of Figure 4.18.

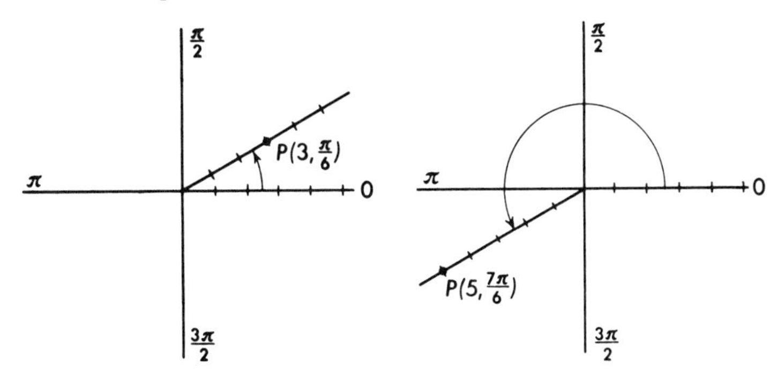


Fig. 4.18 Points in polar coordinates

The quantity $\rho$ differs from the distance $r$, used in the definitions of the trigonometric functions, in that it is a signed number. *Positive* values of $\rho$ indicate distances from the origin *measured along the terminal side of* $\theta$. *Negative* values of $\rho$ are defined as distances measured *along the extension of the terminal side of* $\theta$ *through the pole.* Thus the point $\rho = -4$, $\theta = \pi/6$, or $(-4, \pi/6)$, is shown in Figure 4.19. Note that the points with coordinates $(-4, \pi/6)$,

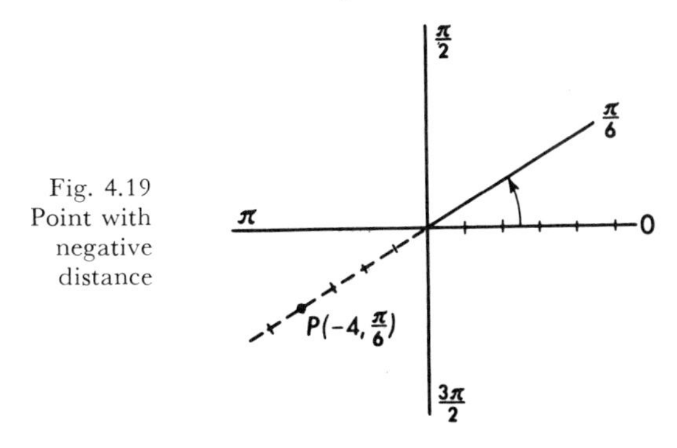


Fig. 4.19 Point with negative distance

$(4, 7\pi/6)$, and $(4, -5\pi/6)$ are identical, so that the polar coordinates of a given point are not unique.

Any equation expressing $\rho$ as a function of $\theta$ restricts the variable points $(\rho, \theta)$ to occupying only those positions for which the values of $\rho$ and $\theta$ satisfy the equation. Hence, by substitution, the graph can be approximated by a sufficient number of points, connected in order of value of $\theta$.

The problem of graphing equations in polar coordinates is not essentially different from that of graphing equations in Cartesian coordinates. The polar equation may usually be solved for $\rho$ explicitly in terms $\theta$, and a table of values of $\rho$ corresponding to selected values of $\theta$ may be computed.

**EXAMPLE**

Trace the locus of the equation $\rho = 4 \sin \theta$.

**Solution:** The loci of simple equations involving $\sin \theta$ or $\cos \theta$ can usually be satisfactorily traced by locating the points corresponding to $\theta = 0$, $\theta = \pi/6$, $\theta = \pi/3$, and so forth (the multiples of $\pi/6$). Other values of $\theta$ may be chosen as it seems helpful to do so. It is suggested that the values of $\rho$ and $\theta$ be tabulated for the relation $\rho = 4 \sin \theta$ as follows:

| $\theta$ | $\sin \theta$ | $\rho\ (= 4 \sin \theta)$ |
|---|---|---|
| $0$ | $0$ | $0$ |
| $\frac{\pi}{6}$ | $0.5$ | $2.0$ |
| $\frac{\pi}{3}$ | $0.87$ | $3.48$ |
| $\frac{\pi}{2}$ | $1$ | $4$ |
| $\frac{2\pi}{3}$ | $0.87$ | $3.48$ |
| $\frac{5\pi}{6}$ | $0.5$ | $2$ |
| $\pi$ | $0$ | $0$ |
| $\frac{7\pi}{6}$ | $-0.5$ | $-2$ |
| $\frac{4\pi}{3}$ | $-0.87$ | $-3.48$ |
| $\frac{3\pi}{2}$ | $-1$ | $-4$ |
| $\frac{5\pi}{3}$ | $-0.87$ | $-3.48$ |
| $\frac{11\pi}{6}$ | $-0.5$ | $-2$ |

It will be observed that when $\theta$ runs through the third and fourth quadrants, $\rho$ is negative, thus locating $P$ in the first and second quadrants, respectively. In fact, for this particular equation the point $P$ *retraces* the path followed as $\theta$ varied from $0$ to $\pi$. By plotting these points and joining them in order with a smooth curve, the locus shown in Figure 4.20 is obtained. This seems to be a

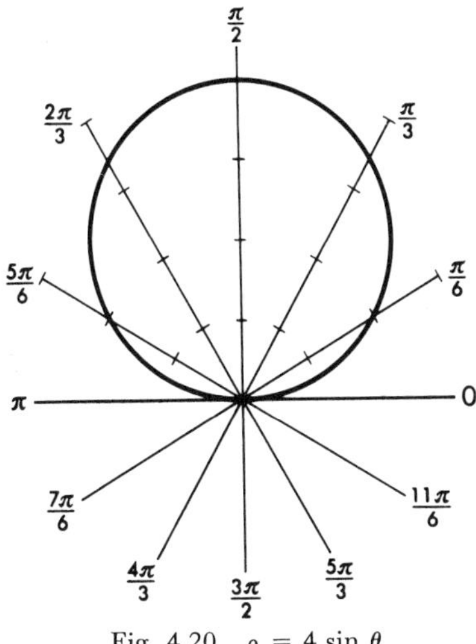


Fig. 4.20 $\rho = 4 \sin \theta$

circle. This can be further checked by trying other values of $\theta$. (See also Exercise 9 of Section 4.8.)

## 4.8 SUGGESTIONS FOR RAPID SKETCHING IN POLAR COORDINATES

In sketching the locus of an equation in polar coordinates such as

$$\rho = 4 \cos 2\theta$$

it is helpful first to sketch the graph of the auxiliary equation

$$y = 4 \cos 2\theta$$

on a rectangular coordinate system with $(\theta, y)$ as coordinates. By using a knowledge of the period ($\pi$) and the amplitude (4), and of the manner of variation of the cosine function, the graph of the auxiliary equation can be readily sketched. It is a simple matter to let the $y$ values for this graph become the $\rho$ values for the sketch in polar coordinates. This method is helpful in sketching any polar curve for which the corresponding auxiliary equation in rectangular coordinates is the easier of the two to sketch.

As an example, continue the consideration of the graph of $\rho = 4 \cos 2\theta$. A sketch of the graph of the auxiliary equation $y = 4 \cos 2\theta$ is shown in Figure 4.21a. Observe that, as $\theta$ increases from 0 to $\pi/4$, $y$ decreases from 4 to 0. Hence, on the polar sketch, Figure 4.21b, $\rho$ decreases from 4 to 0 for the same change in $\theta$. By continuing to follow the variation of $y$ (and, so, of $\rho$)

as $\theta$ increases, develop the polar sketch until a repeating cycle is finished, or until the desired portion of the graph is obtained. Note that any value of $\theta$

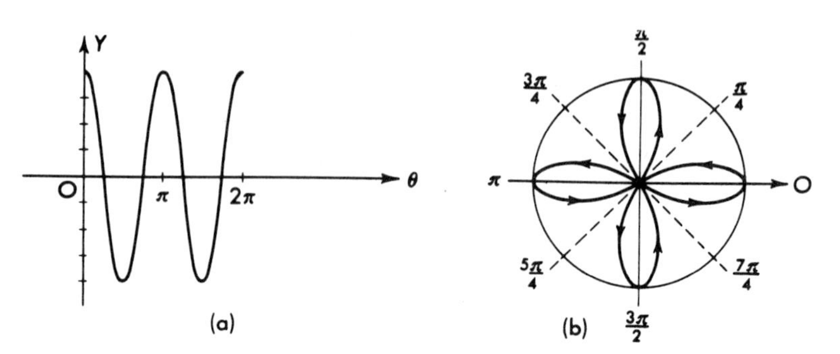


Fig. 4.21 Comparative sketches in rectangular and polar coordinates (a) $y = 4 \cos 2\theta$ (b) $\rho = 4 \cos 2\theta$

for which $\rho = 0$ is the measure of the angle made by the tangent to the curve at the pole with the polar axis.

## *EXERCISES*

1. On one polar coordinate system plot the following points: $A(6, 0)$; $B(4, \pi/4)$; $C(2, 3\pi/4)$; $D(-5, \pi)$; $E(-3, \pi/2)$.
2. For each of the points of Exercise 1, give two different sets of coordinates besides the one given.
3. The point $(4, \pi/6)$ can also be represented by the coordinates $(-4, 7\pi/6)$ and $(-4, -5\pi/6)$. Verify that, in general, this point is represented by $[(-1)^n 4, \pi/6 + n\pi]$, where $n = 0, \pm 1, \pm 2, \pm 3. \ldots$ Give the value of $n$ which, when substituted in the general form, gives each of the three particular representations of the point mentioned above.
4. Sketch each of the following loci in polar coordinates after sketching the auxiliary equation in rectangular coordinates.

   (**a**) $\rho = 6 \cos \theta$
   (**b**) $\rho = 4(1 - \sin \theta)$
   (**c**) $\rho = 2 + 4 \sin \theta$
   (**d**) $\rho = 6 \cos 2\theta$
   (**e**) $\rho = 4 \sin 3\theta$
   (**f**) $\rho = 4|\sin \theta|$

5. Sketch the following loci in polar coordinates.

   (**a**) $\rho = \dfrac{8}{2 - \sin \theta}$
   (**b**) $\rho = 4 \csc \theta$
   (**c**) $\rho = 3 \sin^2 \theta$
   (**d**) $\rho^2 = 4 \cos 2\theta$

   [*Note:* In (**d**), observe that $\rho$ is imaginary when $\cos 2\theta$ is negative, so that for certain values of $\theta$ no points are determined.]

**6.** Verify that the formulas $x = \rho \cos \theta$ and $y = \rho \sin \theta$ can be used to change the equation of a locus from rectangular to polar coordinates.

**7.** Use the formulas of Exercise 6 to change the following equations to polar form. Simplify each result if possible.

(a) $x^2 + y^2 = 16$
(b) $x^2 + y^2 - 2x + 4y = 0$
(c) $y = 4$ [compare with Exercise 5(b) above]
(d) $y = x$
(e) $x^2 + y^2 = 0$

**8.** Without sketching, describe each of the following loci, basing your answer on some observable property of the equation.

(a) $\rho = 8$ (the absence of $\theta$ means "for every value of $\theta$")
(b) $\theta = 2$
(c) $\rho = 0$
(d) $\rho \cos \theta = 1$
(e) $\rho = \dfrac{\theta}{2}$
(f) $\theta = 0$

**9.** *A mettle tester.* Prove that $\rho = 4 \sin \theta$ is the equation of a circle (see the example in Section 4.7). [*Hint:* Prove that for every value of $\theta$ the point $(4 \sin \theta, \theta)$ is at a constant distance from some fixed point which proves to be the center of the circle. Having shown that every point given by the equation is on a circle, you still must show that the equation gives *all* points of the circle.

## *MISCELLANEOUS EXERCISES/CHAPTER 4*

**1.** As the angle $\theta$ increases, determine

(a) the quadrant in which both $\sin \theta$ and $\cos \theta$ are increasing,
(b) which trigonometric function increases as $\theta$ increases through all quadrants,
(c) which trigonometric function decreases as $\theta$ increases through all quadrants,
(d) the quadrant in which both $\sec \theta$ and $\csc \theta$ are increasing,
(e) which trigonometric functions decrease as $\theta$ increases from $90°$ to $270°$,
(f) in which quadrant $\cos \theta$ remains greater than $\sin \theta$.

**2.** Find the maximum value of each of the functions

(a) $3 \sin 2\theta$
(b) $4 \cos \dfrac{\theta}{2}$
(c) $\frac{1}{2} \cos 4\theta$
(d) $-3 \sin \theta$

3. What is the minimum value of each of the functions of Exercise 2?

4. In Chapter 6, we shall be able to show that, for every value of $x$ for which the functions are defined,

(a) $\sin(\pi - x) = \sin x$
(b) $\sin(-x) = -\sin x$
(c) $\sin(\pi + x) = -\sin x$

Use Figure 4.2, the graph of $y = \sin x$, to check that these relations seem true for selected values of $x$.

5. Also to be verified in Chapter 6 are the following relations, which are true for every value of $x$:

(a) $\cos(\pi - x) = -\cos x$
(b) $\cos(-x) = \cos x$
(c) $\cos(\pi + x) = -\cos x$

From a graph of $y = \cos x$, show that these relations seem true for selected values of $x$.

6. On the same rectangular coordinate system sketch the graphs of $y = \tan x$ and $y = \sin(x/2)$ on the interval $0 \leqq x \leqq 2\pi$. From the graph determine the number of values of $x$ on this interval for which $\sin(x/2) = \tan x$. Estimate these values of $x$ to the nearest tenth of $\pi$ (for example, $0.6\pi$ or $1.2\pi$).

7. Repeat Exercise 6 for the graphs of $y = 3\cos x$ and $y = \sin(x/2)$.

8. Repeat Exercise 6 for the graphs of $y = \sin 2x$ and $y = 2\cos x$.

9. Sketch the curve $y = 2\cos(x - \pi/3)$. See if you can do this by a translation of the curve $y = 2\cos x$.

10. Upon one set of axes, sketch the graphs of the polar equations $\rho = 2$ and $\rho = 4\cos\theta$. Shade the region outside the curve $\rho = 2$ and inside the curve $\rho = 4\sin\theta$. What are the points of intersection of the curves? (*Note:* With the aid of calculus, you will be able to find the area of such regions, and to determine other useful things, such as the center of gravity of the region.)

11. Repeat Exercise 10 for the region outside $\rho = 9$ and inside $\rho = 6(1 + \sin\theta)$.

12. Repeat Exercise 10 for the region outside $\rho = 4 + \cos\theta$ and inside $\rho = 3(1 + \cos\theta)$.

13. Find the value of $k$ so that the period of $\sin kx$ is $3\pi/2$.

14. We have taken the domain and the range of each of the six trigonometric functions to be the set of real numbers (with certain numbers deleted to obtain the domain of the functions whose definitions would exclude "division by zero"). In view of

your knowledge of the behavior of the functions, formulate a statement of the numbers included in the domain of each of the functions sine, cosine, and so forth.

**15.** We may say that the range of a function is the set of real numbers, yet have only a limited set of the numbers of the range assigned by the function. For example, the sine function does not pair $\sin \theta$ with 2 for any number $\theta$. The actual subset of the numbers that *are* assigned by a function is called the *image* of the function. For example, the image of the sine function is the interval $[-1,1]$. What is the image of each of the other five trigonometric functions?

# CHAPTER 5. SOLUTION of RIGHT TRIANGLES and APPLICATIONS

## 5.1 INTRODUCTION

The sides and angles of a triangle are called the *parts* of the triangle. A triangle is *determined* if a sufficient number of its parts are known to make it congruent to any other triangle having corresponding parts of equal measure. In particular, a right triangle is determined if (in addition to the right angle) a side and another part (angle or side) are given. The process of finding the unknown parts of a triangle is called *solving the triangle.* This chapter deals with the solution of right triangles and some applications. The solution of the general triangle will be studied in Chapter 8.

In this book a uniform notation for parts of triangles will be employed: Greek letters for the angles, capital Roman letters for the vertices, and small Roman letters for the lengths of the corresponding opposite sides. For example, the angles of triangle $ABC$ will be designated by $\alpha$, $\beta$, and $\gamma$, and the lengths of the opposite sides by $a$, $b$, and $c$, respectively. An angle with vertex $A$ may be called angle $A$ if no ambiguity is involved.

In the solution of triangles we shall distinguish between two types of triangles: (1) mathematical triangles with lengths of sides and measures of angles given exactly, and (2) triangles representing physical counterparts, the sides and angles of which have been determined as approximate numbers by some process of measurement. An example of type (1) is $\triangle ABC$ with $a = 3$, $b = 5$, and $\gamma$ (the angle at $C$) $= 90°$. From the Pythagorean Theorem, we can say that $c = \sqrt{34}$ (exactly). An example of type (2) is the triangle determined by a "vertical" pole and its shadow on level ground. Here the length of the pole, the length of its shadow, and the perpendicularity of the pole, as well as the level state of the ground, are all measured to varying degrees of accuracy. Any computed parts of such a triangle will necessarily be approximate.

Even in solving type (1) triangles, we must remember that values of quantities found in the tables are usually approximate numbers, and if used in computations, lead to approximate solutions, even though the original numbers given as lengths, and so forth were considered as exact.

A reading of Appendix I will be helpful in knowing how much reliance is to be placed upon the accuracy of results computed from approximate numbers. A fair working rule is that (a) *computed* length of sides can be no more accurate than any *given* lengths of sides, and (b) *four* place tables give accuracy sufficient to approximate angles to the *nearest minute*, if it is known that the other numbers used are also accurate to four significant digits. Likewise, *three-figure* accuracy of lengths of sides is matched by accuracy of angle measures to the *nearest 10 minutes*, while *two* figure accuracy of lengths accompanies accuracy of angles to the *nearest degree.*

## 5.2 SOLVING THE RIGHT TRIANGLE

Let the right triangle $ABC$, with the right angle at $C$,* be so placed in relation to a rectangular coordinate system that one of its acute angles, say $\alpha$, is in standard position as shown in Figure 5.1. The assignment rule of the trigonometric functions of $\alpha$ can be restated as follows:

$$\sin\alpha = \frac{a}{c} = \frac{\text{side opposite } \alpha}{\text{hypotenuse}}$$

$$\cos\alpha = \frac{b}{c} = \frac{\text{side adjacent (to } \alpha)}{\text{hypotenuse}}$$

$$\tan\alpha = \frac{a}{b} = \frac{\text{side opposite}}{\text{side adjacent}}$$

$$\cot\alpha = \frac{b}{a} = \frac{\text{side adjacent}}{\text{side opposite}}$$

$$\sec\alpha = \frac{c}{b} = \frac{\text{hypotenuse}}{\text{side adjacent}}$$

$$\csc\alpha = \frac{c}{a} = \frac{\text{hypotenuse}}{\text{side opposite}}$$

Familiarity with these statements makes it possible to express any of the six trigonometric functions of an acute angle of a right triangle in terms of the sides *without regard to the position of the triangle*. These make it possible to choose an equation relating any two of the sides of a right triangle with either of the acute angles. Thus, if one side and an acute angle are known, or if two sides are known, the remaining parts can be calculated. Recall that $\alpha = 90° - \beta$.

---

* Unless otherwise stated, we shall continue to assume that, if $ABC$ is a right triangle, the right angle is at $C$.

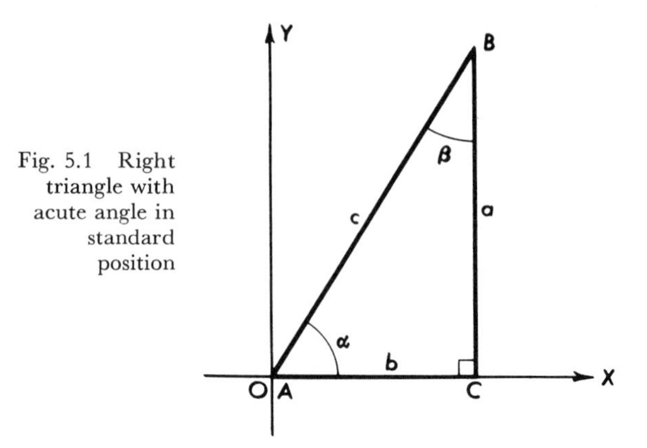


Fig. 5.1 Right triangle with acute angle in standard position

**EXAMPLE**

Solve the right triangle $ABC$, given $a = 6.04$, $\beta = 62° \, 30'$ (Figure 5.2).

**Solution:** Choose the relation

$$\sec \beta = \frac{c}{a}$$

since it involves only one unknown, $c$. Thus, substituting the values of $\beta$ and $a$ and reversing sides gives

$$\frac{c}{6.04} = \sec 62° \, 30'$$

so that

$$c \approx 6.04(2.1657) \approx 13.081$$

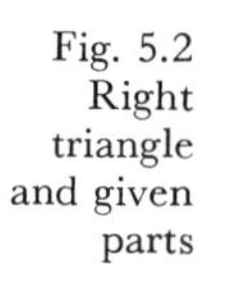

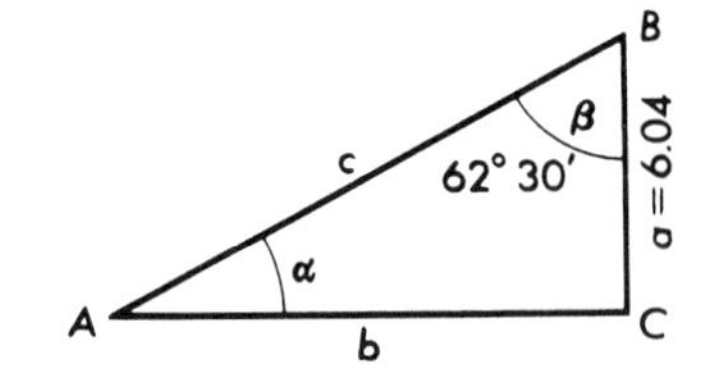


Fig. 5.2 Right triangle and given parts

(Note that $c$ is given to five significant digits, since this is the given accuracy of sec 62° 30′, assuming $a$ exact.)

To find $b$, use

$$\frac{b}{a} = \tan \beta$$

Thus

$$\frac{b}{6.04} \approx 1.921$$

from which

$$b \approx 11.60$$

Finally,

$$\alpha = 90° - \beta$$
$$= 27°\ 30'$$

(*Note:* The equation $\cos \beta = \frac{a}{c}$ could have been used to find $c$, but a long-division operation would have followed. It is best to start an equation with the unknown, if possible.)

As a basis for a check, select the equation

$$\cos \alpha = \frac{b}{c}$$

or, better, the equivalent equation

$$c \cos \alpha = b$$

since it involves the operation of multiplication instead of that of division. We wish to know if the values $\alpha = 27°\ 30'$, $b = 11.60$, and $c = 13.081$ satisfy this equation approximately. Thus we ask the question

"Is $(13.081)(\cos 27°\ 30') \approx 11.60$?"

Since both sides are found to have the value 11.60 when rounded off to four significant digits, the answer is *yes*, and we say that our solution of the triangle checks.

## *EXERCISES*

**1.** Solve each of the following for the parts of right triangle $ABC$ requested to accuracy justified by the given information and the tables. Assume that the angle at $C$ is exactly a right angle.

**(a)** $a = 150$, $\alpha = 64°45'$. Find $c$ and $b$.
**(b)** $\beta = 6°15'$, $c = 140.9$. Find $a$.
**(c)** $b = 12.46$, $c = 17.19$. Find $\alpha$ and $a$.
**(d)** $a = 6.6$, $c = 11$. Find $\beta$.
**(e)** $b = 12$, $\alpha = 60°$. Find $a$, $c$, and $\beta$ (tables not needed).
**(f)** $a = 9$, $c = 18$. Find $\alpha$ and $b$ (tables not needed).
**(g)** $\beta \approx 44°12'$, $a \approx 16.21$. Find $b$ and $c$.
**(h)** $c \approx 40$, $a \approx 19$. Find $\beta$ and $b$.
**(i)** $b \approx 15.3$, $\alpha \approx 23°$. Find $a$.
**(j)** $a \approx 3.0$, $c \approx 5.0$. Find $\beta$ and $b$.

2. A straight highway rises approximately 150 ft in 3000 ft along the highway. Find the angle at which the highway is inclined from the horizontal.
3. A ladder with length approximately 30.0 ft is so placed when its foot is on level ground that it leans against a vertical wall. The ladder makes an angle of 70°30′ with the ground. Approximate the height to which it reaches on the wall.
4. A regular pentagon (five-sided polygon with equal lengths of sides and equal angles) is inscribed in a circle of radius 7. Find the length of a side of the pentagon. (Note: Use of the tables will, of course, result in an approximate answer. It is possible to find an *exact* answer with the aid of Exercise 14, Section 2.6.)
5. The leaning tower of Pisa is approximately 179 ft in height and is approximately 16.5 ft out of plumb. Find the angle at which it deviates from the vertical.
6. Prove that the length of a leg (one of the perpendicular sides) of a right triangle is equal to the product of the length of the hypotenuse and the sine of the angle opposite the leg.
7. State a similar relation expressing the leg in terms of the hypotenuse and the adjacent angle.
8. State a relation expressing one leg in terms of the other leg and the angle (**a**) opposite the first leg, (**b**) adjacent to the first leg.

---

## 5.3 LOGARITHMS OF THE TRIGONOMETRIC FUNCTIONS

Modern computing machines have made obsolete the use of logarithms as an efficient aid in intricate computations involving products, quotients, powers, and roots. However, a knowledge of logarithmic functions *as they express relationships* is of great importance in many applications of mathematics; and one very good way to gain familiarity with logarithms for this purpose is to have some practice in restating equations in logarithmic form and simplifying the results.

The reader is strongly urged to turn to Appendix II and, if needed, refresh his knowledge of logarithms, giving particular attention to Section 7 and its exercises.

Table II enables us to determine the common logarithms of numbers with three significant digits (four, with the aid of interpolation). Table IV provides the logarithms of the trigonometric functions of positive acute angles.

For example, log sin 57° 10′ is found directly in Table IV, instead of by looking up sin 57° 10′ in Table III and then finding the logarithm of this in Table II. It is important to keep in mind the approximate value of the trigonometric function in order to assign the proper characteristic to the logarithm of the function. The sines and cosines of all angles between 0° and 90° are positive numbers that are less than one. Hence log sin $\alpha$ and log cos $\alpha$ have negative characteristics for angles of this range. Similarly, log tan $\alpha$ has a negative characteristic for $0° < \alpha < 45°$, as does log cot $\alpha$ for $45° < \alpha < 90°$. Thus, in using the table of logarithms of the trigonometric functions (Table IV) it is necessary to affix $-10$ to the portion of the logarithm as found in the table whenever the absolute value of the trigonometric function is less than one.

Interpolations are performed with the logarithms of trigonometric functions as in the table of natural functions (Table III) (see Section 2.7).

## 5.4 USE OF LOGARITHMS IN SOLVING RIGHT TRIANGLES

Logarithms are used in the solution of the two right triangles below. The evaluation of missing sides is the main concern of the first solution, while the finding of an unknown angle is a major part of the second.

**EXAMPLE 1**

Solve the right triangle $ABC$, given $a = 85.63$ and $\beta = 43°\ 24'$ (Fig. 5.3).

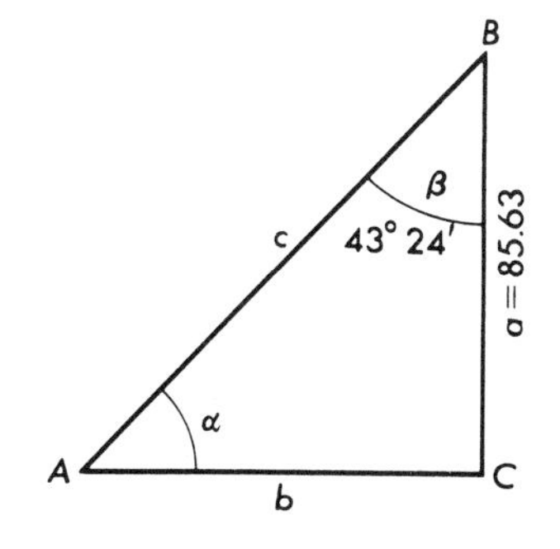


Fig. 5.3 Right triangle with side and acute angle known

**Solution:** Write

$$\cos \beta = \frac{a}{c}$$

If two members are equal, their logarithms are equal. Hence

$$\log \cos \beta = \log a - \log c$$

Thus

$$\begin{aligned} \log c &= \log a - \log \cos \beta \\ &\approx 1.9327 - (9.8613 - 10) \\ &= 2.0714 \end{aligned}$$

Therefore

$$c \approx 117.8$$

Again,

$$\frac{b}{a} = \tan \beta$$

and hence

$$\log b - \log a = \log \tan \beta$$

Hence

$$\begin{aligned} \log b &= \log a + \log \tan \beta \\ &\approx 1.9327 + 9.9757 - 10 \\ &= 1.9084 \end{aligned}$$

Thus

$$b \approx 80.98$$

Finally,

$$\alpha = 90° - 43° \, 24' = 46° \, 36'$$

**EXAMPLE 2**

Solve the right triangle $ABC$ in which $a \approx 17.20$ and $b \approx 24.90$.

**Solution:**

$$\tan \alpha = \frac{a}{b}$$

so that

$$\begin{aligned} \log \tan \alpha &= \log a - \log b \\ &\approx 1.2355 - 1.3962 \\ &= 9.8393 - 10 \end{aligned}$$

From Table IV,

$$\alpha \approx 34° \, 38'$$

Hence

$$\begin{aligned} \beta &\approx 90° - 34° \, 38' \\ &= 55° \, 22' \end{aligned}$$

The computation of $c$ follows the method of Example 1. From

$$\sin \alpha = \frac{a}{c}$$

we obtain

$$\begin{aligned} \log c &= \log a - \log \sin \alpha \\ &\approx 1.2355 - (9.7546 - 10) \\ &= 1.4809 \end{aligned}$$

Hence

$$c \approx 30.26$$

## EXERCISES

1. Use logarithms to solve the right triangle $ABC$ ($C = 90°$) for the side or angle designated. Follow the form of the examples (unless you have a better one) as it is important that you practice making correct statements involving logarithms of numbers. Carefully note the distinction between "$=$" and "$\approx$" and use them as your information dictates.

   (a) Given $a = 14$, $\beta = 81°$, find $b$.
   (b) Given $b = 91.5$, $\alpha = 31°24'$, find $a$.
   (c) Given $a = 27.32$, $\alpha = 37°33'$, find $c$.
   (d) Given $b \approx 24.6$, $\beta \approx 25°20'$, find $a$.
   (e) Given $b = 12.55$, $c = 20.63$, find $\alpha$.
   (f) Given $a \approx 30.1$, $b \approx 17.1$, find $c$.

2. To find $AB$, the distance across a river, a line segment $\overline{AC}$ is laid off perpendicular to $\overline{AB}$. Find the width of the river if $\angle ACB \approx 48°10'$ and $AC \approx 1226$ ft.
3. With the aid of logarithms compute the value of the fraction $f$ where

$$f = \frac{5280(\sin 12°15')(\sin 17°20')}{\sin 5°30'}$$

4. The angle $\theta$ is in standard position and its terminal side passes through the point (57.60, 47.82). Approximate $\theta$ as accurately as possible.
5. Angle $\theta$ is in standard position and its terminal side contains a point $P$ that is 28.9 units distant from the origin. $\theta = 313°30'$. Approximate the coodinates of $P$. (Note that the given numbers are exact; thus any inaccuracy will come from use of the tables, or in rounding off results.)

## 5.5 APPLICATIONS

The solution of triangles finds direct application in the fields of surveying and navigation. In addition, the trigonometric relations among the parts of a triangle are basic in studies of forces and motion, which are so important in many branches of science and engineering. Furthermore, practice in expressing these relationships is valuable background for the analytical aspects of trigonometry needed in today's physics and chemistry. In this section there are several illustrative examples involving the solution of right triangles. Special terms and concepts will be introduced as needed in the form of notes. (The assumption that triangles on the earth's surface are plane triangles still obtains.)

## EXAMPLE 1

An airplane flies 120 mi on a course of 125° 20′. How far is it east and how far is it south of the starting point? [*Note:* A *course* is a directed line of travel. The direction of the course from point $A$ to point $B$ may be given by the *bearing* of the course, which is the angle in degrees measured *clockwise* from north to the direction $\overrightarrow{AB}$. Another description of bearing, used by mariners, gives direction in terms of the number of degrees west or east of the north or south direction. In this system, the bearing of $B$ from $A$ in the example would be given as S 54° 40′ E, that is, 54° 40′ east of south. (See Figure 5.4.)]

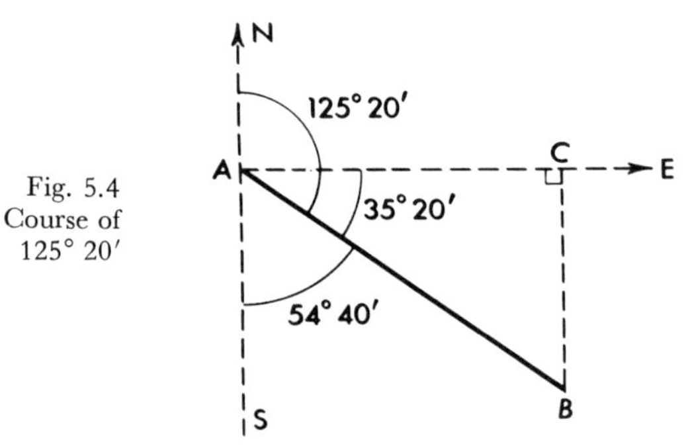


Fig. 5.4 Course of 125° 20′

**Solution:** In Figure 5.4, $A$ is the starting point and $B$ the destination. $AC$ is the eastward distance and $CB$ the southward distance. The given numbers are obviously measurements, and are approximations.

$$\angle BAC \approx 125° \ 20' - 90° = 35° \ 20'$$

Now

$$\frac{AC}{AB} \approx \cos 35° \ 20'$$

Hence

$$\begin{aligned} AC &\approx (120)(0.8158) \\ &\approx 97.9 \text{ miles} \end{aligned}$$

Likewise,

$$\begin{aligned} CB &\approx AB \sin 35° \ 20' \\ &\approx 69.4 \text{ miles} \end{aligned}$$

## EXAMPLE 2

Find the magnitude of the projection of a line segment 16.23 ft. long upon a line making an angle of 37° 14′ with the given segment. [*Note:* The projection of a line segment $AB$ (directed or undirected) upon a line $l$ is the segment

$A'B'$, where $A'$ and $B'$ are the respective points at which perpendiculars from $A$ and $B$ meet $l$. Figure 5.5 suggests how the magnitude of the projection

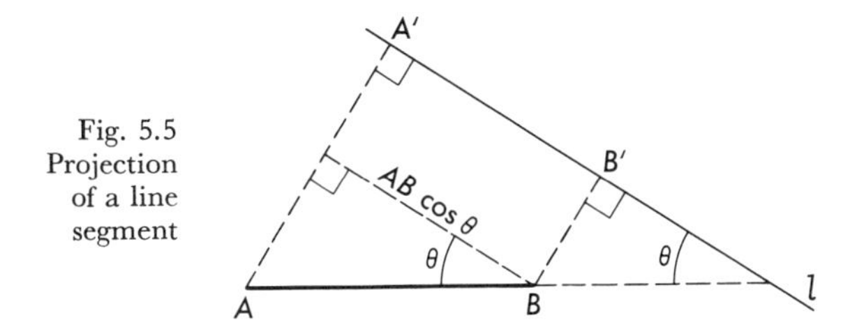


Fig. 5.5 Projection of a line segment

may be found as the side of a right triangle. The result is

$$A'B' = AB \cos \theta$$

where $\theta$ is the positive, acute angle between $\overleftrightarrow{AB}$ and $l$.]

**Solution:** We leave this to the student. *Ans.* 12.92 ft, approximately.

**EXAMPLE 3**

The angle of elevation of the top of a vertical pole is 65°30′, 72 ft from the base of the pole on level ground. Find the height of the pole. (*Note:* If an observer is looking at an object *above* his level, the acute angle between the horizontal direction and the line from the observer to the object is called the *angle of elevation* of the object with respect to the observer. If the object is below the level of the observer, the corresponding angle is the *angle of depression* of the object with respect to the observer.)

**Solution:** If $h$ represents the height of the pole (Figure 5.6), then

$$\frac{h}{72} \approx \tan 65^\circ\, 30'$$

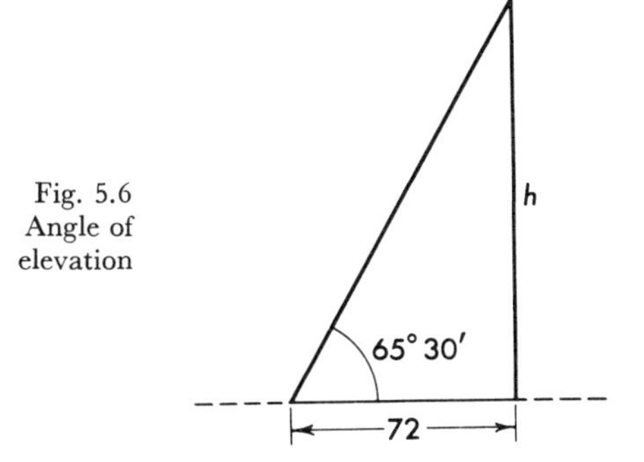


Fig. 5.6 Angle of elevation

Hence

$$h \approx (72)(2.1943)$$
$$\approx 160 \text{ ft}$$

**EXAMPLE 4**

A swimmer is swimming with a speed of 2.0 mph and heads directly across a river having a current speed of 4.0 mph, both speeds approximate. In what direction is the swimmer actually moving, and with what speed? [*Note:* Quantities representing velocities, forces, and so forth, which have *both direction and magnitude* are called vectors. A vector is represented by a directed line segment drawn as an arrow. This arrow, or vector, points in the direction of the vector quantity which it represents, and its length (to a suitable scale) represents the magnitude of the quantity. The total effect, or *resultant*, of two similar vector quantities is given pictorially by the *vector diagonal of the parallelogram* of which the two given vectors form adjacent sides. Two vectors that combine to form a given resultant are called *components* of the resultant vector. It is customary to represent a vector by a letter in boldface type and to represent the length of the vector by the same letter in lightface type.]

**Solution:** In Figure 5.7, vector **s** represents the swimmer's velocity, and vector **c**, the current's velocity. The actual velocity of the swimmer is given

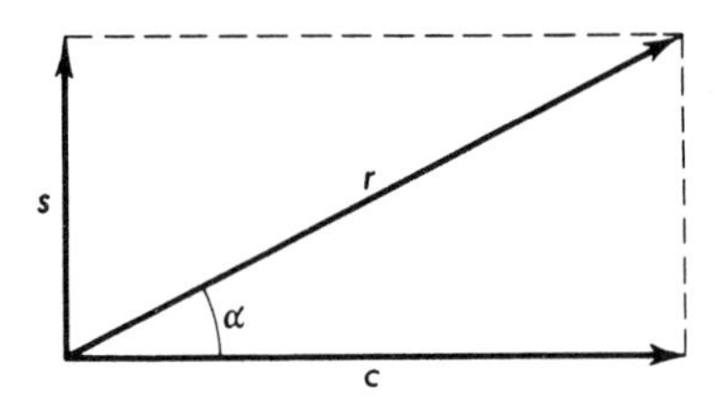


Fig. 5.7 Vectors and their resultant

in magnitude and direction by the resultant vector **r**. From the right-triangle law,

$$r^2 \approx 2^2 + 4^2$$
$$= 20$$

Hence the actual speed of the swimmer is approximately $\sqrt{20}$ mph. Also

$$\tan \alpha \approx \frac{2.0}{4.0} = 0.50$$

Hence

$$\alpha \approx 27°$$

Hence the swimmer moves in a line making an angle of approximately 27° with the bank at a speed of approximately $\sqrt{20}$, or 4.5, mph.

## *EXERCISES*

1. A tower 150 ft in height casts a shadow 50.5 ft long. Find the angle of elevation of the sun.
2. A ship leaves port sailing in the direction of S 40° W. If its speed is 17 knots (17 nautical mph), find how far west and how far south it is from the starting point 2 hr after departure.
3. If the height of a pole is 75.4 ft, find the distance from the foot of the pole an observer must stand so that the angle of elevation of the top is 41°. Assume that his eye level is 5 ft above the ground.
4. Find the horizontal and vertical components of a force of 60 lbs acting at an angle of 37° with the horizontal.
5. A plane flies a course of 270° with a speed of 200 mph. At the end of 2 hr it flies due north for 1 hr at the same speed. Find the distance and direction from the starting point at the end of the 3 hr.

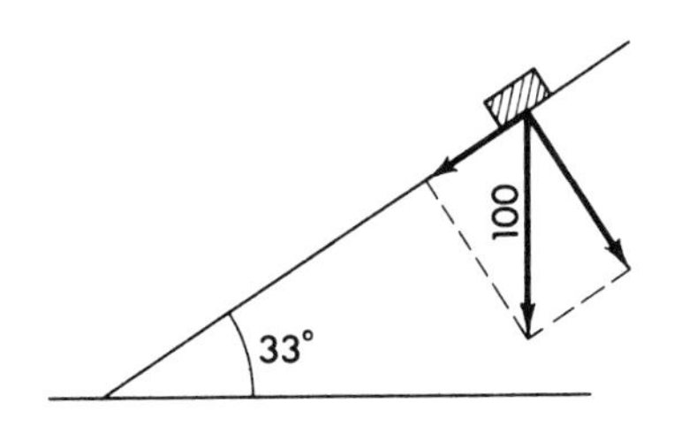


Ex. 5

6. Find the length of a line segment $\overline{AB}$ if the length of its projection on a line at an angle of 34°21′ with it is 25 ft.
7. From a point 120 ft above a level plane the angles of depression of the top and bottom of a tower standing on the plane are 30° and 60° respectively. Find the height of the tower.
8. On the top of a vertical shaft 150 ft tall stands a statue that at a horizontal distance of 225 ft from the foot of the shaft subtends an angle of 2°30′. Find the height of the statue.
9. Given that the area of a triangle is $\frac{1}{2}bh$ where $b$ is the length of any side and $h$ is the length of the altitude of the triangle upon that side, derive a formula for the area $K$ of a triangle in terms of any two sides, $a$ and $b$, and their included angle $C$. (This triangle need not be a right triangle.)
10. Given triangle $ABC$ with $a = 40$, $b = 65$ and $C = 70°20'$, find the area.
11. An airplane is climbing with an airspeed of 200 mph. What is the angle of climb if the vertical rate of climb is 44 fps?

## *MISCELLANEOUS EXERCISES/CHAPTER 5*

1. In the right triangle $ABC$, suppose that $\alpha$ and $b$ are known. Express the missing parts of the triangle in terms of $\alpha$ and $b$.
2. Given angle $\theta$ in standard position, the terminal side passes through $P$ that is $k$ units from the origin. Find the coordinates of $P$ in terms of $\theta$ and $k$.
3. A ship sails in the direction of 97° at a speed of 17 knots. Find its eastward speed.
4. Which of the following are impossible statements? Explain.

   (**a**) $\log \tan \theta = 1.6134$
   (**b**) $\log \sin \theta = 0.3456$
   (**c**) $\log \sec \theta = -1.0000$
   (**d**) $\log \cos \theta = 0.0000$
   (**e**) $\log \cot \theta = 9.4235 - 10$
   (**f**) $\log \csc \theta = 0.4781$

5. A weight of 100 lb is placed on a smooth plane inclined at an angle of 33° with the horizontal, as in the figure. With what force must one push along the incline to just prevent the weight from slipping? (*Hint:* The downward force of 100 lb can be resolved into components in the direction of the incline and perpendicular to it, respectively. A force equal to the component along the incline and directed up the incline is necessary to prevent slipping.)
6. If a segment (see Figure 1.7) of a circle of radius $r$ subtends a central angle of $\theta$ radians, show that $K'$, the area of the segment, is given by the formula

$$K' = \tfrac{1}{2}r^2(\theta - \sin \theta)$$

   (*Hint:* Use the fact that the area of the segment is the area of the sector minus the area of a certain triangle.)
7. A chord of length 10 connects two points on a circle of radius 8. Find the area of the sector formed.
8. A force of 100 lb is directed to the left. What vertical force must act with it so that the resultant has a magnitude of 140 lbs? Find also the angle that the resultant makes with the horizontal.
9. Find the radius of the circle of 30° north latitude on the earth's surface, assuming that the radius of the earth is 4,000 miles and that the earth is spherical.
10. Find the distance along the circle of Exercise 9 between a point in China with longitude 110° east of Greenwich and a point in Algeria of longitude 0°.

**11.** The muzzle velocity of a rifle bullet is 3,500 fps. Find the horizontal and vertical components of this velocity if the rifle is fired at an angle of elevation of 21°.

**12.** A man can paddle a boat 3 mph in still water. He wishes to cross a river whose current is 2 mph. Find the direction he must head the boat to land straight across from his starting point.

**13.** From the top of a tower 170 ft high two points $A$ and $B$ on a straight line through the foot of the tower are observed to have angles of depression of 8° and 6°, respectively. Find the distance $AB$.

**14.** A plane is flying due west at 200 mph, and the wind is blowing from the south at 25 mph. Find the ground speed of the plane and the direction of its flight.

**15.** Find the length of a side of a regular pentagon circumscribed about a circle of radius 15 units.

**16.** Given two planes $\pi$ and $\pi'$ intersecting at an angle $\varphi$ (as in the figure), the projection of the point $P$ in plane $\pi$ is defined as

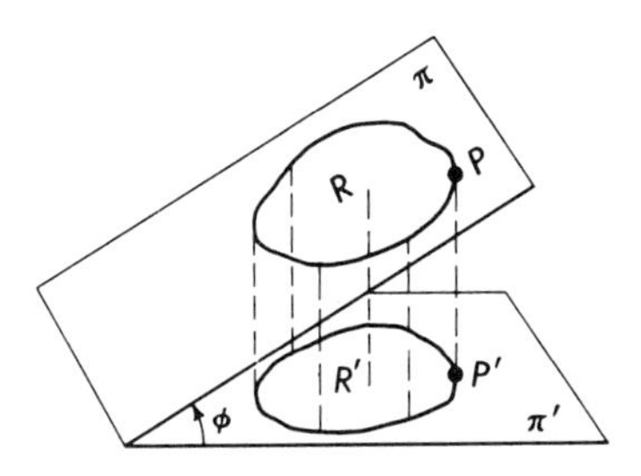


Ex. 16

the point $P'$ at which the perpendicular from $P$ intersects $\pi'$. Similarly, if $R$ is any region (collection of points) of $\pi$, then $R'$, the projection of $R$ onto $\pi'$, is defined as the totality of the projections of the points of $R$ onto $\pi'$. Suppose now that $R$ is a rectangle with one side parallel to the line of intersection of the two planes. Prove that

$$\text{area of } R' = (\text{area of } R) \cdot \cos \varphi$$

**17.** Regarding the sun as a uniformly brilliant disk, determine what fraction of its illumination is in effect when one fourth of its vertical diameter is showing above the horizon.

**18.** A thin belt runs around two circular pulleys with centers 36 in. apart. The radii of the pulleys are 6 in. and 15 in., respectively. Assuming the belt is stretched tightly and does not cross itself between the pulleys, find the length of the belt.

**19.** If $a$ is the length of a side of a regular polygon of $n$ sides, and $R$ is the length of the radius of the circumscribed circle, show that

$$a = 2R \sin \left(\frac{180}{n}\right)^\circ$$

**20.** An observer in a boat at sea notes that the angle of elevation of the top of a lighthouse standing on a cliff is 30°. After he rows 400 yds toward a point directly under the lighthouse, he observes that the angles of elevation of the top and bottom of the tower are 60° and 45° respectively. Find the heights of the cliff and tower without using tables.

# CHAPTER 6. TRIGONOMETRIC IDENTITIES

## 6.1 EQUATIONS AND IDENTITIES

A numerical equation is a statement that two symbols represent the same number. Some examples are:

$$5/10 = 1/2$$
$$1/4 + 1/2 = 3/4$$
$$1 + 3 = 2$$

and, if $x$ is a number,

$$x - 2 = 5$$

and

$$x(x - 2) = x^2 - 2x$$

The first two of these equations are true statements, while the third one is false. If in the last two equations, $x$ is a variable over a set of numbers (that is, may be replaced by any member of the set), then these two equations differ in an important respect. Let us suppose that $x$ is a variable over the set of real numbers $R$. Then the equation

$$x - 2 = 5$$

is *not* true for every number $x$, and, in fact, is true only for $x = 7$. On the other hand, the equation

$$x(x - 2) = x^2 - 2x$$

*is* true for every value of $x$. Thus we may distinguish between two types of equations involving a variable: (1) equations whose *truth set* does not contain all of the numbers of the set over which $x$ varies; and (2) equations whose truth set contains every number of the set for which both sides of the equation have meaning. In passing, we note that the truth set of some equations is the empty set. Examples of these are, if $x \in R$,

$$\sqrt{x^2 + 4} = 1$$

and

$$\sin x = 2$$

Nevertheless, these belong to type (1), since their truth sets certainly do *not* contain all real numbers.

We give a name to type (2) equations by the following definition. An equation that is true for every value of its variable (or variables) for which both sides of the equation are defined is called an *identity*. The triple-bar symbol "$\equiv$" denotes the phrase "is identically equal to" and is often used to call attention that an equation is an identity without having to say this in words. Thus we might write

$$\sqrt{x^2} \equiv |x|$$

if it is understood that $x$ is a variable over $R$. If this is not understood, one had best write:

For every $x$ such that $x \in R$, $\sqrt{x^2} = |x|$.

We can write

$$\frac{2}{x} + \frac{1}{x - 2} \equiv \frac{3x - 4}{x(x - 2)}$$

since this statement is true for every number $x$ except 0 and 2, for which members of the equation are meaningless. (Why meaningless?)

$$(x - y)(x^2 + xy + y^2) = x^3 - y^3$$

is an example of an identity in two variables.

## 6.2 VERIFYING AN IDENTITY

Given an equation in a variable, how can we prove that it is, or is not, an identity? If an equation is not an identity, this is readily shown by finding *one* value of the variable (in the intersection of the domains of the functions represented by the two sides of the equation) for which the equation is not true. For example, the equation

$$\sin \theta = 1 - \cos \theta$$

is not an identity, since it is not a true statement if $\theta$ is replaced by $\pi/6$, since

$$\sin \frac{\pi}{6} = \frac{1}{2}$$

and

$$1 - \cos \frac{\pi}{6} = 1 - \frac{\sqrt{3}}{2}$$

so that

$$\sin\frac{\pi}{6} \neq 1 - \cos\frac{\pi}{6}$$

Of course, this equation is also not true for many other values of $\theta$, but exhibiting one suffices to show that it is not an identity.

To prove that an equation is an identity, it is sufficient to show that the expressions, which the equations states are equal, are both identically equal to a third expression. For instance, the equation $A = B$ is an identity if $A \equiv C$ and $B \equiv C$. Likewise, if we can produce a "chain" of identities, such as $A \equiv C \equiv D \equiv F \equiv B$, then we have $A \equiv B$. We shall accept algebraic simplifications (not involving multiplication by zero or attempted division by zero) as producing an expression identically equal to the given expression. In the following section, we introduce certain basic identities involving trigonometric expressions with the aid of which many other identities can be developed.

**EXAMPLE**

Show that the equation

$$x - [2 - (x - 3)] = 2x - 5$$

is an identity.

**Solution:** By algebraic operations,

$$\begin{aligned} x - [2 - (x - 3)] &\equiv x - (2 - x + 3) \\ &\equiv x - 2 + x - 3 \\ &\equiv 2x - 5 \end{aligned}$$

Hence,

$$x - [2 - (x - 3)] \equiv 2x - 5$$

## *EXERCISES*

In the following, assume that the letters represent real numbers. Determine which of the following equations are identities and which are not. Justify each answer.

1. $\frac{2x}{5} - \frac{3}{2} = \frac{4x - 15}{10}$
2. $(x - y)^2 = x^2 - y^2$
3. $x - \frac{2y + 4}{3} = \frac{3x - 2y - 4}{3}$
4. $(x + y)^3 = x^3 + y^3$
5. $\sqrt{x^2 - y^2} = x - y$
6. $x + y = \sqrt{x^2 + 2xy + y^2}$
7. $\sqrt{x^2 - 2xy + y^2} = |x - y|$
8. $\log_a x - \log_a y = \log_a \frac{x}{y}, a > 0$
9. $(a^{2x})(a^b) = a^{2bx}, a > 0$
10. $\log_a (x + y) = \log_a x + \log_a y, a > 0$

Verify that the following equations are identities by simplifying the more complicated member until it reduces to the other member, that is, by producing a chain of identities. In each case give the set of values of $x$ for which any member is undefined.

**11.** $\dfrac{x^2 - 16}{x^2 - x - 20} = \dfrac{x - 4}{x - 5}$

**12.** $\left(\dfrac{e^x - e^{-x}}{2}\right)^2 - \left(\dfrac{e^x + e^{-x}}{2}\right)^2 = -1$

**13.** $\dfrac{3}{x} + \dfrac{2x}{5} = \dfrac{15 + 2x^2}{5x}$

**14.** $(3x + \sqrt{2})^2 + (3x - \sqrt{2})^2 = 2(9x^2 + 2)$

**15.** $\left(\dfrac{1}{x} - x\right)\left(x + \dfrac{1}{x}\right) = -\left(x^2 - \dfrac{1}{x^2}\right)$

## 6.3 EIGHT FUNDAMENTAL TRIGONOMETRIC IDENTITIES

There are many identities that are repeatedly used to simplify trigonometric expressions. We shall refer to eight of these as the *fundamental trigonometric identities.* They are listed below with some alternative forms on the right of each one. In addition to being able to prove them, the student should memorize them.*

| Identity | | Alternative Forms |
|---|---|---|
| $\sin \theta \equiv \dfrac{1}{\csc \theta}$ | (6.1) | $\csc \theta \equiv \dfrac{1}{\sin \theta}$; $\sin \theta \csc \theta \equiv 1$ |
| $\cos \theta \equiv \dfrac{1}{\sec \theta}$ | (6.2) | $\sec \theta \equiv \dfrac{1}{\cos \theta}$; $\cos \theta \sec \theta \equiv 1$ |
| $\tan \theta \equiv \dfrac{1}{\cot \theta}$ | (6.3) | $\cot \theta \equiv \dfrac{1}{\tan \theta}$; $\tan \theta \cot \theta \equiv 1$ |
| $\tan \theta \equiv \dfrac{\sin \theta}{\cos \theta}$ | (6.4) | $\tan \theta \cos \theta \equiv \sin \theta$ |
| $\cot \theta \equiv \dfrac{\cos \theta}{\sin \theta}$ | (6.5) | $\cot \theta \sin \theta \equiv \cos \theta$ |
| $\sin^2 \theta + \cos^2 \theta \equiv 1$ | (6.6) | $1 - \sin^2 \theta \equiv \cos^2 \theta$; $1 - \cos^2 \theta \equiv \sin^2 \theta$ |
| $1 + \tan^2 \theta \equiv \sec^2 \theta$ | (6.7) | $\sec^2 \theta - \tan^2 \theta \equiv 1$; $\sec^2 \theta - 1 \equiv \tan^2 \theta$ |
| $1 + \cot^2 \theta \equiv \csc^2 \theta$ | (6.8) | $\csc^2 \theta - \cot^2 \theta \equiv 1$; $\csc^2 \theta - 1 \equiv \cot^2 \theta$ |

* In what follows, the symbol "$\sin^2 \theta$" will mean $(\sin \theta)^2$, and similarly for $\cos \theta$, $\tan \theta$, and so forth. In general $\sin^n \theta \equiv (\sin \theta)^n$, *except* for $n = -1$, which case has a different significance to be discussed in Chapter 7.

Each of these identities can be proved by applying the original definitions of the six trigonometric functions. For example, from the fact that $\sin \theta \equiv y/r$ and that $\csc \theta \equiv r/y$, it follows that

$$\begin{aligned}\sin \theta &\equiv \frac{y}{r}\\ &\equiv \frac{1}{\frac{r}{y}}\\ &\equiv \frac{1}{\csc \theta}, \quad \text{if } y \neq 0\end{aligned}$$

which verifies (6.1). If $\theta$ is such an angle (or number) that $y = 0$, then the right-hand member of (6.1) is meaningless. However, the reader will remember that the definition of an identity requires only that the two members be equal for all values of the variable *for which both members are defined.* The verification of (6.2), (6.3), (6.4), and (6.5) in similar fashion is left to the student.

To obtain identity (6.6), make use of the Pythagorean relation $x^2 + y^2 \equiv r^2$. Thus

$$\begin{aligned}\sin^2 \theta + \cos^2 \theta &\equiv \left(\frac{y}{r}\right)^2 + \left(\frac{x}{r}\right)^2\\ &\equiv \frac{y^2 + x^2}{r^2}\\ &\equiv \frac{r^2}{r^2}\\ &\equiv 1\end{aligned}$$

The reader may verify that division of both members of the identity $x^2 + y^2 \equiv r^2$ by $x^2$ (provided $x \neq 0$) results in

$$1 + \tan^2 \theta = \sec^2 \theta$$

and that division by $y^2$ ($y \neq 0$) gives

$$1 + \cot^2 \theta \equiv \csc^2 \theta$$

## 6.4 SIMPLIFYING TRIGONOMETRIC EXPRESSIONS

With the aid of the eight fundamental identities (and, later, of other important ones) many expressions can be reduced to a simpler, or more useful, form. Other identities can also be developed. Some illustrative examples follow. (*Note:* The use of the symbol $\equiv$ in writing equations that are readily recognizable as identities will be discontinued, and this symbol will be reserved for emphasizing the identity relation as it is needed.)

### EXAMPLE 1

Simplify the expression

$$\cos\theta + \tan\theta\sin\theta$$

**Solution:**

$$\begin{aligned}\cos\theta + \tan\theta\sin\theta &= \cos\theta + \frac{\sin\theta}{\cos\theta}\sin\theta \\ &= \frac{\cos\theta}{1} + \frac{\sin^2\theta}{\cos\theta} \\ &= \frac{\cos^2\theta + \sin^2\theta}{\cos\theta} \\ &= \frac{1}{\cos\theta} \\ &= \sec\theta\end{aligned}$$

Thus, by applying formulas (6.4), (6.6), and (6.2) in that order and performing operations of algebra suggested by the form, it is possible to reduce the expression $\cos\theta + \tan\theta\sin\theta$ to the form $\sec\theta$.

### EXAMPLE 2

Show that the equation

$$\tan\alpha + \cot\alpha = \sec\alpha\csc\alpha$$

is an identity.

**Solution:** Consider the left member (with two terms) the more complicated one, and attempt to reduce it to the right member. Thus, using (6.4), (6.5), (6.6), (6.2), and (6.1), respectively, obtain

$$\begin{aligned}\tan\alpha + \cot\alpha &= \frac{\sin\alpha}{\cos\alpha} + \frac{\cos\alpha}{\sin\alpha} \\ &= \frac{\sin^2\alpha + \cos^2\alpha}{\cos\alpha\sin\alpha} \\ &= \frac{1}{\cos\alpha\sin\alpha} \\ &= \sec\alpha\csc\alpha\end{aligned}$$

which verifies the identity.

## *EXERCISES*

Prove that the following are identities by reducing one member to the other or by separately reducing each member to a common form.

1. $\dfrac{\cos\theta}{\sec\theta} + \dfrac{\sin\theta}{\csc\theta} = 1$
2. $\sec^2\theta - \tan^2\theta = \csc^2\theta - \cot^2\theta$
3. $\dfrac{1}{1 - \sin\theta} + \dfrac{1}{1 + \sin\theta} = 2\sec^2\theta$
4. $\dfrac{1 + \cos\theta}{1 - \cos\theta} - \dfrac{1 - \cos\theta}{1 + \cos\theta} = 4\cot\theta\csc\theta$
5. $\dfrac{1 + \cos\theta}{\sin\theta} + \dfrac{\sin\theta}{1 + \cos\theta} = 2\csc\theta$
6. $\dfrac{1}{\tan\theta + \cot\theta} = \dfrac{\sin\theta}{\sec\theta}$
7. $\dfrac{1 + \cos\theta}{\sin\theta} = \dfrac{\sin\theta}{1 - \cos\theta}$
8. $\dfrac{\cot\theta + 1}{\sin\theta + \cos\theta} = \csc\theta$
9. $\dfrac{1 - \cos\theta}{1 + \cos\theta} = (\csc\theta - \cot\theta)^2$
10. $\dfrac{\sec\theta + 1}{\tan\theta} = \dfrac{\tan\theta}{\sec\theta - 1}$
11. $\dfrac{\tan A + \tan B}{\cot A + \cot B} = \tan A \tan B$
12. $\tan\theta\sin\theta\cot^2\theta = \cos\theta$
13. $(\sin\theta + \cos\theta)(\cot\theta + \tan\theta) = \sec\theta + \csc\theta$
14. $(x\sin v + y\cos u)^2 + (x\cos v - y\sin u)^2 = x^2 + y^2$
15. $\dfrac{\sin\theta + 2\sin\theta\cos\theta}{2 + \cos\theta - 2\sin^2\theta} = \tan\theta$
16. $\sin\theta\cos\theta = \dfrac{1}{\tan\theta + \cot\theta}$
17. $\dfrac{\sin\theta}{\cos\theta} + \dfrac{\sec\theta}{\csc\theta} = 2\tan\theta$
18. $\csc^4\theta - \cot^4\theta = \csc^2\theta + \cot^2\theta$
19. $\dfrac{\sin^3\theta + \cos^3\theta}{\sin\theta + \cos\theta} = 1 - \sin\theta\cos\theta$
20. $\dfrac{\tan\theta - \sin\theta\cos\theta}{\sin^2\theta} = \tan\theta$
21. $\dfrac{\sin\theta + \cos\theta\tan\theta}{\tan\theta} = 2\cos\theta$
22. $\dfrac{2\cos^2\theta - \sin^2\theta + 1}{\cos\theta} = 3\cos\theta$
23. $\dfrac{\tan^2\theta}{\sec\theta - 1} = 1 + \sec\theta$
24. $\dfrac{1}{\tan\theta + \sec\theta} = \sec\theta - \tan\theta$

25. $\sin^4 \theta + 2 \sin^2 \theta \cos^2 \theta + \cos^4 \theta = 1$

26. $\dfrac{1 + \cos \theta}{1 - \cos \theta} - \dfrac{1 - \cos \theta}{1 + \cos \theta} = 4 \cot \theta \csc \theta$

27. $(1 - \sin \theta)(1 + \csc \theta) = \cos \theta \cot \theta$

28. $\dfrac{\tan \theta - \sin \theta}{\tan \theta \sin \theta} = \dfrac{\sin \theta \tan \theta}{\tan \theta + \sin \theta}$

29. $\sin \theta\,(1 + \tan \theta) + \cos \theta\,(1 + \cot \theta) = \sec \theta + \csc \theta$

30. $\dfrac{1}{1 + \cos^2 \theta} + \dfrac{1}{1 + \sec^2 \theta} = 1$

31. $\dfrac{\sec \theta}{1 + \tan \theta} = \dfrac{\csc \theta}{1 + \cot \theta}$

In Exercise **32** through **38** assume that $\theta$ is in quadrant I and show that the following identities hold for $\theta$ in this interval.

32. $\cos \theta \sqrt{\sec^2 \theta - 1} = \sin \theta$. Show that this identity holds also for $\theta$ in quadrant III, but not in II or IV.

33. $\sqrt{1 - \sin^2 \theta} = \cos \theta$. In what other quadrant does this hold?

34. $\cos \theta \sqrt{1 + \tan^2 \theta} = 1$. In what other quadrant does this hold?

35. $\sqrt{\dfrac{1 - \cos \theta}{1 + \cos \theta}} = \dfrac{\sin \theta}{1 + \cos \theta}$. In what other quadrant does this hold?

36. If $x = a \sin \theta$, and $a > 0$, show that $\dfrac{x}{\sqrt{a^2 - x^2}} = \tan \theta$, a result useful in calculus, as are the results of Exercises 37 and 38.

37. If $x = a \tan \theta$, $a > 0$, show that $\dfrac{a}{\sqrt{a^2 + x^2}} = \cos \theta$.

38. If $x = a \sec \theta$, $a > 0$, show that $\dfrac{a^2}{x\sqrt{a^2 - x^2}} = \dfrac{\cos^2 \theta}{\sin \theta}$.

At times, the most useful form of a trigonometric expression is not the simplest looking one. For instance, in certain applications of calculus the expression $\dfrac{\cos^2 \theta}{\sin \theta}$ in Exercise 38 would be further transformed as follows:

$$\begin{aligned}\frac{\cos^2 \theta}{\sin \theta} &= \frac{1 - \sin^2 \theta}{\sin \theta}\\ &= \frac{1}{\sin \theta} - \frac{\sin^2 \theta}{\sin \theta}\\ &= \csc \theta - \sin \theta\end{aligned}$$

a form particularly useful in integral calculus.

In Exercises **39** through **45** transform the first expression into the second.

**39.** $\sin^3 \theta$; $\sin \theta - \cos^2 \theta \sin \theta$

**40.** $\cos^5 \theta$; $\cos \theta - 2 \sin^2 \theta \cos \theta + \sin^4 \theta \cos \theta$

**41.** $\sec^4 \theta$; $\sec^2 \theta + \tan^2 \theta \sec^2 \theta$

**42.** $\tan^4 \theta$; $\tan^2 \theta \sec^2 \theta - \sec^2 \theta + 1$

**43.** $\dfrac{1}{1 + \cos \theta}$; $\csc^2 \theta - \csc \theta \cot \theta$

**44.** $\dfrac{\sec^3 \theta}{\tan \theta}$; $\csc \theta + \sec \theta \tan \theta$

**45.** $\dfrac{1 - \cos \theta}{1 + \cos \theta}$; $2 \csc^2 \theta - 2 \csc \theta \cot \theta - 1$

**46.** Prove fundamental identity (6.5) by combining (6.3) and (6.4).

**47.** Prove fundamental identity (6.7) by dividing both members of (6.6) by $\cos^2 \theta$.

**48.** Similarly prove (6.8) from (6.6).

**49.** For each of the eight fundamental identities, state what values of $\theta$ (if any) make the identity meaningless.

---

## 6.5 FUNCTIONS OF NEGATIVES OF ANGLES

Let $\theta$ be any angle, and $-\theta$ its negative. Consider them in standard position as in Figure 6.1. Let $P$ and $P_1$ be the respective points of intersection of the

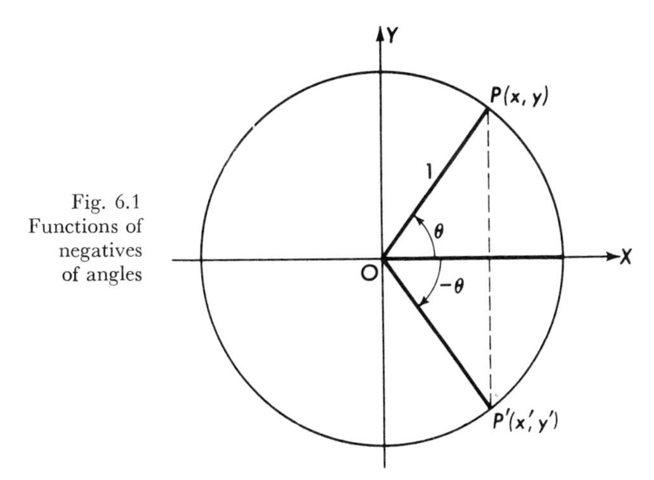


Fig. 6.1 Functions of negatives of angles

terminal sides of $\theta$ and $-\theta$ with the unit circle. Let their rectangular coordinates be denoted by $(x, y)$ and $(x', y')$, respectively. Then for every value of $\theta$,

$$x' = x, \qquad y' = -y$$

But

$$x' = \cos(-\theta), \quad y' = \sin(-\theta)$$
$$x = \cos \theta, \qquad y = \sin \theta$$

Hence, by substitution,

$$\sin(-\theta) = -\sin\theta \tag{6.9}$$

$$\cos(-\theta) = \cos\theta \tag{6.10}$$

Moreover from (6.9), (6.10), and the fact that $\tan\alpha = (\sin\alpha)/(\cos\alpha)$, for all values of $\alpha$ for which the functions are defined,

$$\tan(-\theta) = -\tan\theta \tag{6.11}$$

The student can establish corresponding formulas for $\cot(-\theta)$, $\sec(-\theta)$, and $\csc(-\theta)$.

## *EXERCISES*

1. Express each of the following in terms of functions of positive acute angle and find their values.

   (a) $\cos(-60°)$
   (b) $\sin(-210°)$
   (c) $\tan(-150°)$
   (d) $\cot(-300°)$
   (e) $\sec(-225°)$
   (f) $\csc(-420°)$
   (g) $\cos(-120°)$
   (h) $\sin(-135°)$

2. Find the formulas for $\cot(-\theta)$, $\sec(-\theta)$, and $\csc(-\theta)$, corresponding to numbers (6.9), (6.10), and (6.11).
3. Show that the following are identities.

   (a) $\sin\theta\sin(-\theta) - \cos\theta\cos(-\theta) = -1$
   (b) $\sin(-\theta)\sec\theta = \tan(-\theta)$
   (c) $\sin^2(-\theta) + \cos^2(-\theta) = 1$
   (d) $(1 + \cos\theta)(1 - \cos(-\theta)) = \sin^2\theta$
   (e) $\sec(-\theta)(\sin\theta) = \tan\theta$
   (f) $\sin 210° + \cos 60° = 0$
   (g) $\dfrac{\csc(-150°)}{\csc 30°} = -1$
   (h) $\dfrac{\sin(-30°)}{\cos(-30°)} = -\tan 30°$

4. A function $f$ is called *odd* if $f(-x) = -f(x)$. For instance, $x^3$ is an odd function, since $(-x)^3 = -x^3$. Likewise, $f$ is *even* if $f(-x) = f(x)$. $x^2$ is an even function because $(-x)^2 = x^2$. Determine whether each of the following functions is odd, even, or neither. Justify your answer.

   (a) $\sin x$
   (b) $\cos x$
   (c) $\sin x + \cos x$
   (d) $\sin x\cos x$
   (e) $\sin x + \tan x$
   (f) $1 + \cos x$
   (g) $\sin^2 2x$
   (h) $\sin x\tan x$
   (i) $1 + \sin x$

## 6.6 INTRODUCTION TO OTHER IMPORTANT IDENTITIES

In addition to the eight fundamental relations of the trigonometric functions, there is another important group of identities that expresses such trigonometric expressions as $\sin(\theta + \varphi)$, $\tan(\theta - \varphi)$, $\cos 2\theta$, and so forth, in terms of functions of the single angles. These formulas find frequent use in the applications of trigonometry.

The student is warned to avoid such errors as that of assuming

$$\cos(\theta + \varphi)$$

to be equivalent to $\cos\theta + \cos\varphi$. This fallacy should not tempt anyone who remembers that $\cos(\theta + \varphi)$ is read "the cosine *of* $(\theta + \varphi)$" and not "the cosine *times* $(\theta + \varphi)$." The fact that $\cos(\theta + \varphi)$ is not, in general, equal to $\cos\theta + \cos\varphi$ is readily shown by substituting particular values of $\theta$ and $\varphi$, say $\theta = 60°$, $\varphi = 30°$, in the two expressions. Thus

$$\cos(60° + 30°) = \cos 90° = 0$$

But

$$\cos 60° + \cos 30° = \frac{1}{2} + \frac{\sqrt{3}}{2} > 0$$

Since 0 is not equal to $\frac{1}{2} + \frac{\sqrt{3}}{2}$, it is evident that

$$\cos(60° + 30°) \neq \cos 60° + \cos 30°$$

The following articles state and verify the correct formulas for expressing $\cos(\theta + \varphi)$, $\cos(\theta - \varphi)$, $\sin(\theta + \varphi)$, $\sin(\theta - \varphi)$, $\tan(\theta + \varphi)$, and

$$\tan(\theta - \varphi)$$

in terms of functions of $\theta$ and $\varphi$ alone. The formulas are frequently called the *addition formulas* of trigonometry. The *double-angle formulas* for $\sin 2\theta$, $\cos 2\theta$, and $\tan 2\theta$, and the *half-angle formulas* for $\sin\frac{\theta}{2}$, $\cos\frac{\theta}{2}$, and $\tan\frac{\theta}{2}$ will also be developed.

## 6.7 THE COSINE ADDITION FORMULAS

Given any two angles $\theta$ and $\varphi$, the following formulas are true:

$$\cos(\theta + \varphi) \equiv \cos\theta\cos\varphi - \sin\theta\sin\varphi \qquad \textbf{(6.12)}$$
$$\cos(\theta - \varphi) \equiv \cos\theta\cos\varphi + \sin\theta\sin\varphi \qquad \textbf{(6.13)}$$

**PROOF OF (6.12):**

Place angles $\theta$, $\theta + \varphi$, and $-\varphi$ in standard position, let $P_1(x_1, y_1)$, $P_2(x_2, y_2)$, and $P_3(x_3, y_3)$ be the respective points of intersection of their terminal sides with the unit circle, and also let $P_4(x_4, y_4)$ be the point $(1, 0)$. (See Figure 6.2.) Then $P_2P_4 = P_1P_3$, because they are chords of the unit circle determined by central angles, both of measure $|\theta + \varphi|$. Therefore, by the formula for the distance between two points,

$$\sqrt{(x_2 - x_4)^2 + (y_2 - y_4)^2} = \sqrt{(x_1 - x_3)^2 + (y_1 - y_3)^2}$$

But

$$\begin{aligned} x_1 &= \cos\theta, & y_1 &= \sin\theta \\ x_2 &= \cos(\theta + \varphi) & y_2 &= \sin(\theta + \varphi) \\ x_3 &= \cos(-\varphi) = \cos\varphi, & y_3 &= \sin(-\varphi) = -\sin\varphi \\ x_4 &= 1, & y_4 &= 0 \end{aligned}$$

Hence the equation of distances becomes

$$\sqrt{[\cos(\theta + \varphi) - 1]^2 + \sin^2(\theta + \varphi)}$$
$$= \sqrt{(\cos\theta - \cos\varphi)^2 + (\sin\theta + \sin\varphi)^2}$$

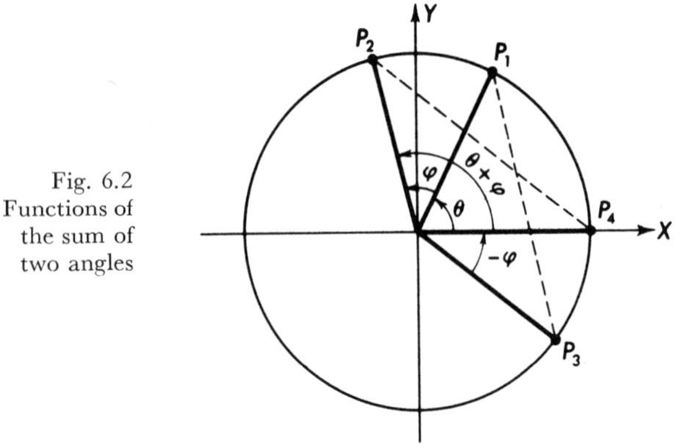


Fig. 6.2
Functions of
the sum of
two angles

With both sides squared and terms collected, this becomes, with the aid of (6.6),

$$2 - 2\cos(\theta + \varphi) = 2 - 2\cos\theta\cos\varphi + 2\sin\theta\sin\varphi$$

which simplifies to the required formula,

$$\cos(\theta + \varphi) = \cos\theta\cos\varphi - \sin\theta\sin\varphi$$

**PROOF OF (6.13)**

Formula (6.12) has been verified for any angles $\theta$ and $\varphi$, positive or negative. Hence, by means of (6.12),

$$\cos(\theta - \varphi) = \cos[\theta + (-\varphi)] = \cos\theta\cos(-\varphi) - \sin\theta\sin(-\varphi)$$

But

$$\cos(-\varphi) = \cos\varphi$$

and

$$\sin(-\varphi) = -\sin\varphi$$

Hence

$$\cos(\theta - \varphi) = \cos\theta\cos\varphi + \sin\theta\sin\varphi$$

### EXAMPLE 1

Find the exact value of cos 75°. (The value found in the table is only *approximately* correct.)

**Solution:** Express 75° in terms of angles whose trigonometric functions are known exactly. Thus

$$75^\circ = 45^\circ + 30^\circ$$

Therefore

$$\begin{aligned}\cos 75^\circ &= \cos(45^\circ + 30^\circ)\\ &= \cos 45^\circ \cos 30^\circ - \sin 45^\circ \sin 30^\circ\\ &= \left(\frac{1}{\sqrt{2}}\right)\left(\frac{\sqrt{3}}{2}\right) - \left(\frac{1}{\sqrt{2}}\right)\left(\frac{1}{2}\right)\\ &= \frac{\sqrt{6}-\sqrt{2}}{4}\end{aligned}$$

(*Note:* This exact answer must be left in radical form. Why?)

### EXAMPLE 2

Simplify the expression $\cos(\pi - \theta)$.

**Solution:** From (6.13),

$$\begin{aligned}\cos(\pi - \theta) &= \cos\pi\cos\theta + \sin\pi\sin\theta\\ &= (-1)\cos\theta + (0)\sin\theta\\ &= -\cos\theta\end{aligned}$$

### EXAMPLE 3

Express $\cos 4\alpha - \cos 6\alpha$ in the form of a product.

**Solution:**

$$4\alpha = 5\alpha - \alpha, \text{ and } 6\alpha = 5\alpha + \alpha$$

Hence

$$\begin{aligned}\cos 4\alpha - \cos 6\alpha &= \cos(5\alpha - \alpha) - \cos(5\alpha + \alpha)\\ &= \cos 5\alpha\cos\alpha + \sin 5\alpha\sin\alpha\\ &\quad - (\cos 5\alpha\cos\alpha - \sin 5\alpha\sin\alpha)\\ &= 2\sin 5\alpha\sin\alpha\end{aligned}$$

## EXERCISES

1. By using $15° = 45° - 30°$, show that $\cos 15° = (\sqrt{6} + \sqrt{2})/4$.
2. Find the exact value of $\cos 105°$.
3. Simplify $\cos 5x \cos x - \sin 5x \sin x$.
4. Simplify $\cos 2x \cos x + \sin 2x \sin x$.
5. Simplify $\cos 70° \cos 20° - \sin 70° \sin 20°$. (Can be done in two ways.)
6. Simplify $\cos 65° \cos 35° + \sin 65° \sin 35°$.
7. By using (6.12) and (6.13), show that

   (a) $\cos(\pi/2 - \theta) = \sin\theta$
   (b) $\cos(\pi/2 + \theta) = -\sin\theta$
   (c) $\cos(\pi + \theta) = -\cos\theta$
   (d) $\cos(\pi - \theta) = -\cos\theta$
   (e) $\cos(3\pi/2 - \theta) = -\sin\theta$
   (f) $\cos(3\pi/2 + \theta) = \sin\theta$
   (g) $\cos(2\pi + \theta) = \cos\theta$
   (h) $\cos(2\pi - \theta) = \cos\theta$

8. Express $\cos 75° + \cos 45°$ as a product by using $75 = 60 + 15$ and $45 = 60 - 15$, and simplify the result.
9. Similarly simplify each of the following:

   (a) $\cos 60° - \cos 30°$
   (b) $\cos 100° + \cos 80°$
   (c) $\cos(x - y) - \cos(x + y)$

10. Express each of the following as a product:

    (a) $\cos 3x - \cos x$
    (b) $\cos 5x + \cos 3x$
    (c) $\cos x + \cos 3x + \cos 5x$
    (d) $\cos 75° + \sin 75°$. Think!

11. Show that $\cos 140° + \cos 100° + \cos 20° = 0$, exactly.
12. Verify each of the following identities:

    (a) $2\cos(\theta - \pi/3) = \cos\theta + \sqrt{3}\sin\theta$
    (b) $\cos 8x \cos 2x + \sin 8x \sin 2x = \cos 5x \cos x - \sin 5x \sin x$
    (c) $\cos(\theta + \theta) = \cos^2\theta - \sin^2\theta$
    (d) $\cos^2(\theta + 60°) + \cos^2(\theta - 60°) + \cos^2\theta = 3/2$

## 6.8 GENERALIZATION OF THE RELATIONS $\cos\left(\frac{\pi}{2} - \theta\right) \equiv \sin\theta$ and $\sin\left(\frac{\pi}{2} - \theta\right) \equiv \cos\theta$

In Section 2.4, it was noted that, for any acute angle $\theta$,

$$\cos\left(\frac{\pi}{2} - \theta\right) \equiv \sin\theta \tag{6.14}$$

$$\sin\left(\frac{\pi}{2} - \theta\right) \equiv \cos\theta \tag{6.15}$$

It is now possible to show that these relations hold for any value of $\theta$.

By (6.13),

$$\cos\left(\frac{\pi}{2} - \theta\right) = \cos\frac{\pi}{2}\cos\theta + \sin\frac{\pi}{2}\sin\theta$$
$$= \sin\theta,$$

which proves (6.14) for the general case.
Since (6.14) is now known to be true for any value of $\theta$, by substituting

$$\theta = \frac{\pi}{2} - \alpha$$

in it, there results

$$\cos\left[\frac{\pi}{2} - \left(\frac{\pi}{2} - \alpha\right)\right] = \sin\left(\frac{\pi}{2} - \alpha\right)$$

or

$$\cos\alpha = \sin\left(\frac{\pi}{2} - \alpha\right)$$

This holds for any value of $\alpha$, proving (6.15) in general.

## 6.9 THE SINE ADDITION FORMULAS

The following two formulas are true for any values of the angles $\theta$ and $\varphi$:

$$\sin(\theta + \varphi) \equiv \sin\theta\cos\varphi + \cos\theta\sin\varphi \qquad \textbf{(6.16)}$$
$$\sin(\theta - \varphi) \equiv \sin\theta\cos\varphi - \cos\theta\sin\varphi \qquad \textbf{(6.17)}$$

**PROOF OF (6.16)**

From (6.14), followed by (6.13), (6.14), and (6.15),

$$\begin{aligned}\sin(\theta + \varphi) &= \cos\left[\frac{\pi}{2} - (\theta + \varphi)\right]\\ &= \cos\left[\left(\frac{\pi}{2} - \theta\right) - \varphi\right]\\ &= \cos\left(\frac{\pi}{2} - \theta\right)\cos\varphi + \sin\left(\frac{\pi}{2} - \theta\right)\sin\varphi\\ &= \sin\theta\cos\varphi + \cos\theta\sin\varphi\end{aligned}$$

**PROOF OF (6.17)**

Note that

$$\sin(\theta - \varphi) = \sin(\theta + [-\varphi])$$

Hence, from (6.16), (6.9), and (6.10)

$$\begin{aligned}\sin(\theta - \varphi) &= \sin\theta\cos(-\varphi) + \cos\theta\sin(-\varphi)\\ &= \sin\theta\cos\varphi - \cos\theta\sin\varphi\end{aligned}$$

## EXERCISES

**1.** By using $15° = 45° - 30°$ show that the exact value of

$$\sin 15° = \frac{\sqrt{6} - \sqrt{2}}{4}$$

**2.** Find the exact value of $\sin 75°$.
**3.** Simplify $\sin 40° \cos 20° + \cos 40° \sin 20°$.
**4.** Simplify $\sin 75° \cos 15° - \sin 15° \cos 75°$.
**5.** By using (6.16) and (6.17) show that

(**a**) $\sin (90° - x) = \cos x$
(**b**) $\sin (90° + x) = \cos x$
(**c**) $\sin (180° + x) = -\sin x$
(**d**) $\sin (180° - x) = \sin x$
(**e**) $\sin (270° - x) = -\cos x$
(**f**) $\sin (270° + x) = -\cos x$
(**g**) $\sin (360° + x) = \sin x$
(**h**) $\sin (360° - x) = -\sin x$

**6.** Simplify $\sin 3x \cos x - \sin x \cos 3x$.
**7.** Simplify $\sin \frac{3}{2}x \cos \frac{1}{2}x + \sin \frac{1}{2}x \cos \frac{3}{2}x$.
**8.** Express $\sin 80° - \sin 60°$ as a product by using $80° = 70° + 10°$, and $60° = 70° - 10°$.
**9.** Express $\sin 80° + \sin 60°$ as a product.
**10.** Simplify $\sin (x + y) + \sin (x - y)$.
**11.** Find the value of $\sin (A + B)$ and $\sin (A - B)$ when

(**a**) $\cos A = \frac{3}{5}$, $A$ in quadrant IV; $\sin B = \frac{4}{5}$, $B$ in quadrant II.
(**b**) $\cos A = \frac{3}{5}$, $A$ in quadrant I; $\sin B = -\frac{4}{5}$, $B$ in quadrant III.
(**c**) $\cos A = -\frac{3}{5}$, $A$ in quadrant II; $\sin B = -\frac{4}{5}$, $B$ in quadrant IV.
(**d**) $\cos A = -\frac{3}{5}$, $A$ in quadrant III; $\sin B = \frac{4}{5}$, $B$ in quadrant I.

**12.** Prove that the following are identities.

(**a**) $\sin (60° + x) + \sin (60° - x) = \sqrt{3} \cos x$
(**b**) $\sin (60° - x) - \sin (60° + x) = -\sin x$

**13.** By using Exercise 8 as a guide, express the following as a product.

(**a**) $\sin 70° + \sin 10°$
(**b**) $\sin 70° - \sin 10°$
(**c**) $\sin 3x + \sin x$
(**d**) $\sin 4x - \sin 2x$

**14.** Verify that

$$\sin (110° - \theta) \cos (160° + \theta) + \cos (110° - \theta) \sin (160° + \theta) = -1$$

**15.** Verify that

$$\frac{\cos (x + y) + \cos (x - y)}{\sin (x + y) + \sin (x - y)} = \cot x$$

**16.** Verify that

$$\frac{\sin 75° + \sin 15°}{\sin 75° - \sin 15°} = \sqrt{3}$$

**17.** Verify that

$$\sin (60° + x) - \sin (60° - x) = \sin x$$

**18.** Show that

$$\sin (x + y + z) = \sin x \cos y \cos z + \cos x \sin y \cos z + \cos x \cos y \sin z - \sin x \sin y \sin z$$

**19.** Find a similar formula for $\cos (x + y + z)$

---

## 6.10 THE TANGENT ADDITION FORMULAS

The following two formulas are valid for all values of $\theta$ and $\varphi$, except when $\theta$, $\varphi$, $\theta + \varphi$, or $\theta - \varphi$ are odd multiples of $\frac{\pi}{2}$:

$$\tan (\theta + \varphi) \equiv \frac{\tan \theta + \tan \varphi}{1 - \tan \theta \tan \varphi} \qquad \textbf{(6.18)}$$

$$\tan (\theta - \varphi) \equiv \frac{\tan \theta - \tan \varphi}{1 + \tan \theta \tan \varphi} \qquad \textbf{(6.19)}$$

**PROOF OF (6.18)**

From (6.4), (6.16), and (6.12)

$$\tan (\theta + \varphi) = \frac{\sin (\theta + \varphi)}{\cos (\theta + \varphi)}$$

$$= \frac{\sin \theta \cos \varphi + \cos \theta \sin \varphi}{\cos \theta \cos \varphi - \sin \theta \sin \varphi}$$

Dividing the numerator and denominator of this fraction by $\cos \theta \cos \varphi$,

$$\tan (\theta + \varphi) = \frac{\dfrac{\sin \theta \cos \varphi}{\cos \theta \cos \varphi} + \dfrac{\cos \theta \sin \varphi}{\cos \theta \cos \varphi}}{\dfrac{\cos \theta \cos \varphi}{\cos \theta \cos \varphi} - \dfrac{\sin \theta \sin \varphi}{\cos \theta \cos \varphi}}$$

Simplifying,

$$\tan (\theta + \varphi) = \frac{\tan \theta + \tan \varphi}{1 - \tan \theta \tan \varphi}$$

The student should verify (6.19).

## EXERCISES

1. Find $\tan 15°$ by using $15° = 45° - 30°$.
2. Find $\tan 75°$.
3. Find the exact value of $\tan 150°$ by using $150° = 75° + 75°$. Why not use $150° = 90° + 60°$?
4. Given $\tan x = 3/4$, $\tan y = 5/12$, find the value of $\tan(x + y)$ and of $\tan(x - y)$.
5. Prove $\tan(\pi/4 - x)\tan(\pi/4 + x) = 1$.
6. Simplify $\dfrac{\tan(x+y) - \tan y}{1 + \tan(x+y)\tan y}$. (Look before you leap.)
7. Prove the following identities.

   (a) $\tan(\pi/2 - \theta) = \cot\theta$ (Use $\pi/2 = \pi/4 + \pi/4$)
   (b) $\tan(\pi/2 + \theta) = -\cot\theta$
   (c) $\tan(\pi + \theta) = \tan\theta$
   (d) $\tan(\pi - \theta) = -\tan\theta$

8. Show that
$$\frac{\sin(x+y) + \sin(x-y)}{\cos(x+y) + \cos(x-y)} = \tan x$$
9. Show that
$$\cot(x+y) = \frac{\cot x \cot y - 1}{\cot x + \cot y}$$
10. Show that
$$\cot(x-y) = \frac{\cot x \cot y + 1}{\cot y - \cot x}$$
11. Given $\tan x = a$, $\tan y = -1/a$, show that $\tan(x - y)$ is not defined. What does this show about $(x - y)$?
12. Prove $\tan(\theta + 60°) + \tan(\theta - 60°) = \dfrac{8\cot\theta}{\cot^2\theta - 3}$.
13. *A (lucky) mettle tester.* Show that
$$\tan 70° - \tan 20° = 2\tan 40° + 4\tan 10°$$
14. Prove that
$$\tan(x+y+z) = \frac{\tan z + \tan y + \tan z - \tan x \tan y \tan z}{1 - \tan x \tan y - \tan x \tan z - \tan y \tan z}$$

## 6.11 THE DOUBLE-ANGLE FORMULAS

If $\varphi = \theta$, the trigonometric functions of $\theta + \varphi$ become the functions of $2\theta$. Thus, from (6.16),

$$\begin{aligned}\sin 2\theta = \sin(\theta + \theta) &= \sin\theta\cos\theta + \cos\theta\sin\theta \\ &= 2\sin\theta\cos\theta\end{aligned}$$

Hence

$$\sin 2\theta \equiv 2 \sin \theta \cos \theta \qquad \textbf{(6.20)}$$

Similarly,

$$\cos 2\theta \equiv \cos^2 \theta - \sin^2 \theta \qquad \textbf{(6.21)}$$

Substituting $\cos^2 \theta = 1 - \sin^2 \theta$ in (6.21) gives

$$\cos 2\theta \equiv 1 - 2 \sin^2 \theta \qquad \textbf{(6.21')}$$

Substituting $\sin^2 \theta = 1 - \cos^2 \theta$ in (6.21) gives

$$\cos 2\theta \equiv 2 \cos^2 \theta - 1 \qquad \textbf{(6.21'')}$$

It is left to the student to verify that

$$\tan 2\theta \equiv \frac{2 \tan \theta}{1 - \tan^2 \theta} \qquad \textbf{(6.22)}$$

## *EXERCISES*

1. Given $\cos 15° \approx 0.97$, find $\cos 30°$ approximately by (6.21″) and check with the value in Table III.
2. Given, $\sin 30° = \frac{1}{2}$, $\cos 30° = \sqrt{3}/2$, find $\sin 60°$.
3. Given $\sin 75° \approx 0.97$, approximate $\cos 150°$.
4. Given $\tan 30° = 1/\sqrt{3}$, use (6.22) to find $\tan 60°$).
5. Given $\tan \theta = \frac{1}{3}$, find $\tan 2\theta$.
6. Assuming that the $\theta$ of Exercise 5 is in quadrant I, find $\sin 2\theta$ and $\cos 2\theta$.
7. If the $\theta$ of Exercise 5 is in quadrant III, find $\sin 2\theta$, $\cos 2\theta$.
8. Given $\sin x = \frac{1}{3}$, with $x$ in quadrant II, and $\cos y = \frac{2}{3}$, with $y$ in quadrant I, find the value of $\sin (2x + y)$.
9. Given $\cos 2\theta = \frac{3}{4}$, find $\cos 4\theta$.

Prove that equations 10 – 15 are identities.

10. $\cos 4x = 1 - 8 \sin^2 x \cos^2 x$
11. $\dfrac{1 + \cos 2x}{\sin 2x} = \cot x$
12. $(\sin A + \cos A)^2 = 1 + \sin 2A$
13. $\tan \theta + \cot \theta = 2 \csc 2\theta$
14. $\dfrac{1 - \cos 2x}{\sin 2x} = \tan x$
15. $\dfrac{1 + \sin 2\theta + \cos 2\theta}{1 + \sin 2\theta - \cos 2\theta} = \cot \theta$
16. In integral calculus, it is often necessary to express even powers of $\sin x$ and $\cos x$ in terms of first degree expressions. With the aid of (6.21′) and (6.21″), show that

$$\sin^2 x = \frac{1 - \cos 2x}{2}$$

and that

$$\cos^2 x = \frac{1 + \cos 2x}{2}$$

**17.** As in Exercise 16, express $\sin^2 3x$ and $\cos^2 3x$ in terms of first degree functions.

**18.** Express $\sin^4 x$ and $\cos^4 x$ in terms of first degree functions. *Hint:* Write $\sin^4 x = (\sin^2 x)^2$ and use the formulas of Exercise 16.

## 6.12 THE HALF-ANGLE FORMULAS

Solving (6.21′) for $\sin^2 \theta$ gives (see Exercise 16 of the preceding section)

$$\sin^2 \theta = \frac{1 - \cos 2\theta}{2}$$

Since this is true for every value of $\theta$, it is true, in particular, for $\theta = \varphi/2$, so that

$$\sin^2 \frac{\varphi}{2} \equiv \frac{1 - \cos \varphi}{2} \qquad \textbf{(6.23)}$$

[*Note:* Taking square roots, we have *either* $\sin \varphi/2 = \sqrt{\dfrac{1 - \cos \varphi}{2}}$ or

$$\sin \frac{\varphi}{2} = -\sqrt{\frac{1 - \cos \varphi}{2}}$$

the sign being chosen according to the quadrant in which $\varphi/2$ terminates, as the definition of the sine requires. For example, if $\varphi = 210°$, then $\varphi/2 = 105°$ and

$$\begin{aligned}\sin 105° &= \sqrt{\frac{1 - \cos 210°}{2}} \\ &= \sqrt{\frac{1 - \left(-\frac{\sqrt{3}}{2}\right)}{2}} \\ &= \frac{\sqrt{2 + \sqrt{3}}}{2}\end{aligned}$$

But, if $\varphi = 390°$, so that $\varphi/2 = 195°$, then (noting that $\sin 195°$ is negative)

$$\begin{aligned}\sin 195° &= -\sqrt{\frac{1 - \cos 390°}{2}} \\ &= -\sqrt{\frac{1 - \frac{\sqrt{3}}{2}}{2}} \\ &= -\frac{\sqrt{2 - \sqrt{3}}}{2}\end{aligned}$$]

A formula for $\cos^2 \varphi/2$ may similarly be derived from (6.21″). It is

$$\cos^2 \frac{\varphi}{2} \equiv \frac{1 + \cos \varphi}{2} \qquad \textbf{(6.24)}$$

As in the above note, a formula for $\cos \varphi/2$ is obtained by extracting the square root of each member of (6.24). Again the choice of sign for the radical is determined by the quadrant in which $\varphi/2$ terminates.

A convenient identity for $\tan \theta/2$ is obtained by applying, in order, (6.4), (6.20), (6.24), and a little algebra, as follows. For every $\theta$ such that $\theta/2$ is in the doman of the tangent function,

$$\begin{aligned}\tan \tfrac{1}{2}\theta &= \frac{\sin \tfrac{1}{2}\theta}{\cos \tfrac{1}{2}\theta} \\ &= \frac{2 \sin \tfrac{1}{2}\theta \cos \tfrac{1}{2}\theta}{2 \cos^2 \tfrac{1}{2}\theta} \\ &= \frac{\sin \theta}{1 + \cos \theta}\end{aligned}$$

Hence,

$$\tan \tfrac{1}{2}\theta \equiv \frac{\sin \theta}{1 + \cos \theta}$$

**EXAMPLE**

Evaluate $\cos 3\pi/4$ by using the fact that $\cos 3\pi/2 = 0$.

**Solution:** Using (6.24) with $\phi = 3\pi/2$, and thus $\phi/2 = 3\pi/4$, we write

$$\begin{aligned}\cos^2 \frac{3\pi}{4} &= \frac{1 + \cos 3\pi/2}{2} \\ &= \tfrac{1}{2}\end{aligned}$$

Hence,

$$\cos \frac{3\pi}{4} = \pm \frac{1}{\sqrt{2}}$$

But $3\pi/4$ is the measure of an angle of the second quadrant, hence the choice of sign must be the negative one. Thus

$$\cos \frac{3\pi}{4} = -\frac{1}{\sqrt{2}}$$

*Comment:* This result was known as well as the given information. The example was used to illustrate the significance of the sign in this connection.

## *EXERCISES*

1. Find the value of $\cos 15°$ by using $15 = 30/2$. Repeat for $\sin 15°$ and $\tan 15°$.
2. In Exercise 1 of Section 6.7, you were asked to show that $\cos 15° = (\sqrt{6} + \sqrt{2})/4$ by using $15 = 45 - 30$. Show that this and the answer to Exercise 1 above are equal. *Hint:* Under what conditions does the equation $a^2 = b^2$ imply that $a = b$?
3. Find the exact values of $\sin 22°30'$, $\cos 22°30'$, and $\tan 22°30'$. Use the fact that $22\frac{1}{2} = 45/2$. Replace your values by decimal approximations and check with Table III.
4. Find the value of $\tan (x/2)$ if $\sin x = \frac{4}{5}$ and $\cos x$ is positive. Also evaluate $\sin (x/2)$ and $\cos (x/2)$.
5. Given $\cos \theta = 24/25$ and $\theta$ is a positive acute angle, find $\sin (\theta/2)$, $\cos (\theta/2)$, and $\tan (\theta/2)$.
6. Given $\sin \theta = -\frac{3}{5}$ and $\theta$ in the interval $(-\pi, -\pi/2)$, find $\sin (\theta/2)$, $\cos (\theta/2)$, and $\tan (\theta/2)$.
7. Express each of the following in terms of the first power of the cosine of double the given angle by using (6.23) or (6.24).

   **(a)** $\cos^2 3x$ **(c)** $\sin^2 2x$ **(e)** $\cos^2 100\pi$
   **(b)** $\sin^2 (5\theta/2)$ **(d)** $\cos^2 (x - y/2)$ **(f)** $\sin^2 x^2$

8. Verify that each of the following is an identity.

   **(a)** $\tan (x/2) = \dfrac{1 - \cos x}{\sin x}$

   **(b)** $\tan (x/2) = \csc x - \cot x$

   **(c)** $\sin y \cot (y/2) = 2 \cos^2 (y/2)$

   **(d)** $\dfrac{1 - \tan x/2}{1 + \tan x/2} = \dfrac{\cos x}{1 + \sin x}$

   **(e)** $\dfrac{1 + \sin x - \cos x}{1 + \sin x + \cos x} = \tan (x/2)$. *Hint:* Use $\sin x = 2 \sin \frac{1}{2}x \cos \frac{1}{2}x$, $1 - \cos x = 2 \sin^2 (x/2)$, and $1 + \cos x = 2 \cos^2 (x/2)$.

9. Certain developments in analytic geometry involve finding $\sin \alpha$ and $\cos \alpha$ when given $\tan 2\alpha$. Find $\sin \alpha$ and $\cos \alpha$ if $\tan 2\alpha = 7/24$ and $2\alpha$ is a positive acute angle.

## 6.13 THE SUM OR DIFFERENCE OF SINES AND OF COSINES

How to express the sum of sines and of cosines in the form of products for special cases has been shown earlier. For example,

$$\begin{aligned}\sin 5x + \sin 3x &= \sin (4x + x) + \sin (4x - x)\\ &= \sin 4x \cos x + \cos 4x \sin x\\ &\qquad\qquad + \sin 4x \cos x - \cos 4x \sin x\\ &= 2 \sin 4x \cos x\end{aligned}$$

This method of factoring, based on the addition formulas, is recommended if it is desired to keep the number of memorized formulas to a minimum. There are times, however, when the quoting of a formula giving the factors is convenient. Therefore these formulas are derived below; the method illustrated above is used. Thus

$$\begin{aligned}\sin \theta + \varphi &= \sin \left(\frac{\theta + \varphi}{2} + \frac{\theta - \varphi}{2}\right) + \sin \left(\frac{\theta + \varphi}{2} - \frac{\theta - \varphi}{2}\right)\\ &= \sin \left(\frac{\theta + \varphi}{2}\right) \cos \left(\frac{\theta - \varphi}{2}\right) + \cos \left(\frac{\theta + \varphi}{2}\right) \sin \left(\frac{\theta - \varphi}{2}\right)\\ &\qquad + \sin \left(\frac{\theta + \varphi}{2}\right) \cos \left(\frac{\theta - \varphi}{2}\right) - \cos \left(\frac{\theta + \varphi}{2}\right) \sin\end{aligned}$$

Hence

$$\sin \theta + \sin \varphi \equiv 2 \sin \left(\frac{\theta + \varphi}{2}\right) \cos \left(\frac{\theta - \varphi}{2}\right) \tag{6.26}$$

It is left to the student to verify in the same manner that

$$\sin \theta - \sin \varphi \equiv 2 \cos \left(\frac{\theta + \varphi}{2}\right) \sin \left(\frac{\theta - \varphi}{2}\right) \tag{6.27}$$

$$\cos \theta + \cos \varphi \equiv 2 \cos \left(\frac{\theta + \varphi}{2}\right) \cos \left(\frac{\theta - \varphi}{2}\right) \tag{6.28}$$

$$\cos \theta - \cos \varphi \equiv -2 \sin \left(\frac{\theta + \varphi}{2}\right) \sin \left(\frac{\theta - \varphi}{2}\right) \tag{6.29}$$

## *EXERCISES*

**1.** Express as the product of two trigonometric functions the following.

(**a**) $\sin 80° + \sin 40°$
(**b**) $\sin 80° - \sin 40°$
(**c**) $\cos 80° + \cos 40°$
(**d**) $\cos 80° - \cos 40°$
(**e**) $\sin 3\theta - \sin \theta$
(**f**) $\cos 4\theta - \cos \theta$

**2.** Show that $\cos 50° + \cos 70° = \cos 10°$.

**3.** Show that $\sin 50° + \sin 70° = \sqrt{3} \sin 80°$.

**4.** Show that $\dfrac{\cos 75° + \cos 15°}{\cos 75° - \cos 15°} = -\sqrt{3}$.

**5.** Prove that the following are identities.

(**a**) $\sin 5x + \sin 3x = 2 \sin 4x \cos x$
(**b**) $\cos 4x + \cos 2x = 2 \cos 3x \cos x$

(c) $\sin 5x - \sin 3x = 2 \cos 4x \sin x$
(d) $\cos 2\theta - \cos 4\theta = 2 \sin 3\theta \sin \theta$
(e) $\sin (60° - y) - \sin (60° + y) = -\sin y$
(f) $2 \sin 4x \sin x = \cos 3x - \cos 5x$

6. The identity

$$\sin (u + h) - \sin u \equiv 2 \cos \left(u + \frac{h}{2}\right) \sin \frac{h}{2}$$

is fundamental in most calculus texts. Verify it.

---

## *MISCELLANEOUS EXERCISES/CHAPTER 6*

**PART I. "QUICKIE" EXERCISES** Each of the following may be evaluated or greatly simplified with one, or at most, two mental steps. You should be able to write the answer down without scratchwork.

1. $2 \sin 75° \cos 75°$
2. $\cos 75° \cos 15° + \sin 75° \sin 15°$
3. $\dfrac{\tan (\pi/4 + \theta) - \tan \theta}{1 + \tan (\pi/4 + \theta) \tan \theta}$
4. $2 \cos^2 4x - 1$
5. $\sin 6x \cos 5x - \cos 6x \sin 5x$
6. $\sin (A + B) + \sin (A - B)$
7. $\cos^2 \dfrac{3x}{2} - \sin^2 \dfrac{3x}{2}$
8. $\cos^2 3x + \sin^2 3x$
9. $\tan \theta \cos \theta \csc \theta$
10. $\dfrac{2 \tan 2y}{1 - \tan^2 2y}$
11. $1 - 2 \sin^2 \frac{1}{2}\theta$
12. $\cos (x + y) \cos (x - y) - \sin (x + y) \sin (x - y)$
13. $\sec^2 5\theta - 1$
14. $\sec \theta(1 - \sin^2 \theta)$
15. $1 + \dfrac{\cot \alpha}{\tan \alpha}$
16. $\cos^2 \theta - \frac{1}{2}$
17. $\cos (A + B) + \cos (A - B)$
18. $\cos (\pi + \theta)$
19. $\sin \theta \cos (\theta - 30°) - \cos \theta \sin (\theta - 30°)$
20. $\sin \frac{1}{2}x \cos \frac{1}{2}x$

**PART 2. WRITTEN EXERCISES** Verify that numbers 1–20 are identities.

1. $\sin 2\theta = \dfrac{2 \tan \theta}{1 + \tan^2 \theta}$
2. $\dfrac{\sin 3\theta}{\sin \theta} - \dfrac{\cos 3\theta}{\cos \theta} = 2$
3. $\dfrac{2}{1 + \cos 2\theta} = \sec^2 \theta$
4. $\tan \theta + \cot \theta = \dfrac{2}{\sin 2\theta}$
5. $\cot \theta - \cot 2\theta = \csc 2\theta$
6. $\sin^2 3\theta - \cos^2 3\theta = \cos (\pi - 6\theta)$
7. $\dfrac{\sin x + \sin y}{\cos x + \cos y} = \tan \tfrac{1}{2}(x + y)$
8. $\dfrac{\sin x - \sin y}{\cos x + \cos y} = \tan \tfrac{1}{2}(x - y)$
9. $\dfrac{\sin x + \sin y}{\sin (x + y)} = \dfrac{\sin (x - y)}{\sin x - \sin y}$
10. $\dfrac{1 + \tan \dfrac{x}{2}}{1 - \tan \dfrac{x}{2}} = \sec x + \tan x$
11. $\sin 3\theta = 3 \sin \theta - 4 \sin^3 \theta$
12. $2 \sin 2\theta \cos \theta - \sin 3\theta = \sin \theta$
13. $\cos 3\theta = 4 \cos^3 \theta - 3 \cos \theta$
14. $\tan 3\theta = \dfrac{3 \tan \theta - \tan^3 \theta}{1 - 3 \tan^2 \theta}$
15. $\dfrac{\cos x + \sin x}{\cos x - \sin x} - \dfrac{\cos x - \sin x}{\cos x + \sin x} = 2 \tan 2x$
16. $\sin x + \sin 3x + \sin 5x + \sin 7x = 4 \cos x \cos 2x \sin 4x$
17. $\dfrac{\cot x + \tan y}{\tan y} = \dfrac{\cos (x - y)}{\sin x \sin y}$
18. $\cos 5\theta = 5 \cos \theta - 20 \cos^3 \theta + 16 \cos^5 \theta$
19. $\dfrac{\cos^3 x - \cos x + \sin x}{\cos x} = \tan x - \sin^2 x.$
20. $\sin^3 \theta + \cos^3 \theta + \sin \theta \cos^2 \theta + \sin^2 \theta \cos \theta = \sin \theta + \cos \theta$

In Exercises 21–25, assume that $A$, $B$, and $C$ are angles of a triangle, so that $A + B + C = 180°$. Verify that each equation is true.

21. $\sin C = \sin (A + B)$ and $\cos C = -\cos (A + B)$, with similar equations for $\sin B$, $\cos B$, $\sin A$, and $\cos A$.
22. $\sin \tfrac{1}{2}(A + B) = \cos \tfrac{1}{2}C$ and $\cos \tfrac{1}{2}(A + B) = \sin \tfrac{1}{2}C$

**23.** $\sin A + \sin B + \sin C = 4 \cos \frac{1}{2}A \cos \frac{1}{2}B \cos \frac{1}{2}C$

**24.** $\tan A + \tan B + \tan C = \tan A \tan B \tan C$

**25.** $\cos A + \cos B + \cos C = 1 + 4 \sin \frac{1}{2}A \sin \frac{1}{2}B \sin \frac{1}{2}C$

Verify equations 26–30 without tables.

**26.** $\dfrac{1 + \sin 45^\circ}{\cos 45^\circ} = \tan 67^\circ 30'$

**27.** $\cos 20^\circ - \cos 80^\circ - \cos 40^\circ = 0$

**28.** $\cos 70^\circ + \sin 40^\circ = \cos 10^\circ$

**29.** $\csc 10^\circ - \sqrt{3} \sec 10^\circ = 4$

**30.** *A mettle tester.* $\sin^2 18^\circ + \cos^2 36^\circ = \frac{3}{4}$. *Note:* This can be solved mechanically with the aid of Exercises 14, Section 2.6. However a more sporting and instructive solution is obtained by using the following hints. Note that $5 \times 36 = 180$. Hence $3(36^\circ) = 180^\circ - 2(36^\circ)$. This means that $\sin 3(36^\circ) = \sin 2(36^\circ)$. Use Exercise 11 above and formula (6.20) and proceed "on your own."

**31.** Sketch the graph of $y = \sqrt{2} \cos (x - \pi/4)$. Why should this look so much like the graph of $y = \sin x + \cos x$?

**32.** Two students solved a problem in differential equations. One obtained the answer

$$\frac{3 - 3 \cos \theta - \sin \theta}{1 - \cos \theta + 3 \sin \theta}$$

The other's was

$$\frac{5 - 4 \cos \theta - 3 \sin \theta}{4 \sin \theta - 3 \cos \theta}$$

Show that if one was right the other was, also.

**33.** Six students worked a problem in integral calculus. They obtained the respective results,

$$\frac{1}{2} \log \frac{1 + \sin x}{1 - \sin x}; \quad \log (\sec x + \tan x); \quad \log \frac{\cos x}{1 - \sin x}; \quad \log \frac{1 + \sin x}{\cos x};$$

$-\log (\sec x - \tan x)$; and $\log \tan (\pi/4 + x/2)$. It was known that $x$ was in the interval $(0, \pi/2)$. Show that if one answer was correct, so were the others.

**34.** By setting $\theta = \alpha + \beta$ and $\varphi = \alpha - \beta$ in formulas (6.27), (6.28), and (6.29), derive the following identities.

$$\begin{aligned}
\sin \alpha \cos \beta &= \tfrac{1}{2}[\sin (\alpha + \beta) + \sin (\alpha - \beta)] \\
\cos \alpha \cos \beta &= \tfrac{1}{2}[\cos (\alpha + \beta) + \cos (\alpha - \beta)] \\
\sin \alpha \sin \beta &= \tfrac{1}{2}[\cos (\alpha - \beta) - \cos (\alpha + \beta)]
\end{aligned}$$

**SUMMARY OF FORMULAS**

$$\sin\theta = \frac{1}{\csc\theta} \tag{6.1}^*$$

$$\cos\theta = \frac{1}{\sec\theta} \tag{6.2}^*$$

$$\tan\theta = \frac{1}{\cot\theta} \tag{6.3}^*$$

$$\tan\theta = \frac{\sin\theta}{\cos\theta} \tag{6.4}^*$$

$$\cot\theta = \frac{\cos\theta}{\sin\theta} \tag{6.5}$$

$$\sin^2\theta + \cos^2\theta = 1 \tag{6.6}^*$$

$$1 + \tan^2\theta = \sec^2\theta \tag{6.7}$$

$$1 + \cot^2\theta = \csc^2\theta \tag{6.8}$$

$$\sin(-\theta) = -\sin\theta \tag{6.9}^*$$

$$\cos(-\theta) = \cos\theta \tag{6.10}^*$$

$$\tan(-\theta) = -\tan\theta \tag{6.11}$$

$$\cos(\theta+\varphi) = \cos\theta\cos\varphi - \sin\theta\sin\varphi \tag{6.12}^*$$

$$\cos(\theta-\varphi) = \cos\theta\cos\varphi + \sin\theta\sin\varphi \tag{6.13}$$

$$\cos\left(\frac{\pi}{2}-\theta\right) = \sin\theta \tag{6.14}$$

$$\sin\left(\frac{\pi}{2}-\theta\right) = \cos\theta \tag{6.15}$$

$$\sin(\theta+\varphi) = \sin\theta\cos\varphi + \cos\theta\sin\varphi \tag{6.16}^*$$

$$\sin(\theta-\varphi) = \sin\theta\cos\varphi - \cos\theta\sin\varphi \tag{6.17}$$

$$\tan(\theta+\varphi) = \frac{\tan\theta+\tan\varphi}{1-\tan\theta\tan\varphi} \tag{6.18}$$

$$\tan(\theta-\varphi) = \frac{\tan\theta-\tan\varphi}{1+\tan\theta\tan\varphi} \tag{6.19}$$

$$\sin 2\theta = 2\sin\theta\cos\theta \tag{6.20}^*$$

$$\cos 2\theta = \cos^2\theta - \sin^2\theta \tag{6.21}^*$$

$$\cos 2\theta = 1 - 2\sin^2\theta \tag{6.21'}$$

$$\cos 2\theta = 2\cos^2\theta - 1 \tag{6.21''}$$

$$\tan 2\theta = \frac{2\tan\theta}{1-\tan^2\theta} \tag{6.22}$$

$$\sin^2\frac{\theta}{2} \equiv \frac{1-\cos\theta}{2} \tag{6.23}^*$$

$$\cos^2\frac{\theta}{2} \equiv \frac{1+\cos\theta}{2} \tag{6.24}^*$$

$$\tan\frac{\theta}{2} \equiv \frac{1-\cos\theta}{\sin\theta} \equiv \frac{\sin\theta}{1+\cos\theta} \tag{6.25}$$

$$\sin \theta + \sin \varphi = 2 \sin \left(\frac{\theta + \varphi}{2}\right) \cos \left(\frac{\theta - \varphi}{2}\right) \quad \textbf{(6.26)}$$

$$\sin \theta - \sin \varphi = 2 \cos \left(\frac{\theta + \varphi}{2}\right) \sin \left(\frac{\theta - \varphi}{2}\right) \quad \textbf{(6.27)}$$

$$\cos \theta + \cos \varphi = 2 \cos \left(\frac{\theta + \varphi}{2}\right) \cos \left(\frac{\theta - \varphi}{2}\right) \quad \textbf{(6.28)}$$

$$\cos \theta - \cos \varphi = -2 \sin \left(\frac{\theta + \varphi}{2}\right) \sin \left(\frac{\theta - \varphi}{2}\right) \quad \textbf{(6.29)}$$

* These identities form a group of key formulas which may be regarded as the minimum essential list. The others can be derived as corollaries. For example, (6.5) is obtained by noting that from (6.3) the cotangent is the reciprocal of the tangent. This, with (6.4), gives (6.5).

Formula (6.7) can be derived from (6.6) by dividing both members by $\cos^2 \theta$ and then using (6.4) and (6.2).

The use of formulas (6.26) to (6.29) can usually be avoided by employing the method of factoring presented in the text.

The student may correctly observe that this list can be further reduced by noting that (6.20) and (6.21) can be obtained from (6.16) and (6.12) by replacing $\varphi$ with $\theta$. However, the ready recognition of (6.20) and (6.21) is so essential that they have been retained in the minimum list.

As an exercise, the unmarked formulas should be derived from the marked ones.

# CHAPTER 7. TRIGONOMETRIC EQUATIONS. INVERSE TRIGONOMETRIC FUNCTIONS

## 7.1 TRIGONOMETRIC EQUATIONS

An equation involving trigonometric functions of a variable over a set of angles (or over a set of numbers) is called a *trigonometric equation*. The set of values of the variable for which the equation is satisfied (that is, for which the equation is a true statement) is called the *solution set*, or *truth set* of the equation. The elements of the solution set are called *roots* of the equation. To *solve* an equation is to find its roots. The process of solving an equation is often called the *solution* of the equation, just as one speaks of the solution of any problem.

We have previously observed, in effect, that an equation whose solution set includes all values of the variable for which any functions involved are defined (meaning that the solution set is the intersection of the domains of all functions involved) is called an identity.

In the following section the solutions of certain simple trigonometric equations will be discussed. This will be followed in later sections with suggestions for reducing the solution of more complicated types to steps involving only the simpler forms.

## 7.2 SOLUTION OF THE EQUATIONS $\sin x = k$, $\cos x = k$, AND SO FORTH

In Section 4.1, we observed that if $\theta$ is an angle in standard position and $P$ is the intersection of the terminal side of $\theta$ and the unit circle with center at the origin, then $\sin \theta = y$ and $\cos \theta = x$, where $y$ and $x$ are the ordinate and abscissa of $P$, respectively. (See Figure 4.1) Now if $k$ is a number such that

$-1 < k < 1$, we see that as $\theta$ assumes all values on the semiclosed interval $0 \leqq \theta < 2\pi$, $\sin \theta\ (= y)$ takes on the value $k$ for exactly two values of $\theta$. This is made clear by noting that the set of all points whose ordinates are $k$ is a straight line, parallel to the $x$ axis, which intersects the $y$ axis at $(0, k)$. The intersection set of this line with the unit circle consists of exactly two points for each specified $k$. If $k = 1$, or $k = -1$, the intersection is a single point (point of tangency), so that $\sin \theta = 1$ but for one value of $\theta$ on $(\theta \leqq \theta < 2\pi)$, namely $\theta = \pi/2$. Likewise $\sin \theta = -1$ for precisely $\theta = 3\pi/2$; for the prescribed $\theta$ interval.

From the above facts we conclude that the equation

$$\sin \theta = k, \ (-1 \leqq k < 1)$$

has exactly two roots on the interval $[0, 2\pi)$ except that there is just one root on the interval for $k = 1$ and for $k = -1$. Now since $\sin \theta$ is periodic, with period $2\pi$, we see that if $\theta_1$ is a root of $\sin \theta = k$, then so is $\theta = \theta_1 + 2\pi$, and in general, so is $\theta_1 + n \cdot 2\pi$, where $n$ is any integer.* This means that we obtain the complete solution set of the equation $\sin \theta = k$ if we find all roots over any interval covering a complete period of the function and then add integral multiples of the period to obtain all other roots.

Similar statements apply to solving the equations

$$\cos \theta = k, \ (-1 \leqq k \leqq 1),$$
$$\tan \theta = k, \text{ for every number } k$$

and so forth. It must be remembered that the period of $\tan \theta$ is $\pi$, so that if $\theta_1$ is a root of $\tan \theta = k$, then $\theta_1 + n\pi$ is also a root for $n = 0, \pm 1, \pm 2, \ldots$

**EXAMPLE 1**

Solve the equation

$$\sin x = \tfrac{1}{2}$$

**Solution:** From our practice work in previous chapters, we should readily observe that $\sin \pi/6 = \frac{1}{2}$ and that $\sin 5\pi/6 = \frac{1}{2}$. We now know that these are the only roots on the interval $[0, 2\pi)$. Hence *all* roots of the equation are given by the formulas

$$x = \frac{\pi}{6} + n \cdot 2\pi \quad \text{and} \quad x = \frac{5\pi}{6} + n \cdot 2\pi \ (n, \text{ any integer})$$

We note that $x$ may be regarded as a variable over the set of real numbers, or as a variable over the set of all angles, the interpretation to be chosen according to the context in which the equation arises. If $x$ is a variable angle, degree

* Recall that the set of integers is the set $I = \{\ldots, -3, -2, -1, 0, 1, 2, 3, \ldots\}$.

measure may be employed, and the solution set of our equation is given by the formulas

$$x = 30° + n \cdot 360° \text{ and } x = 150° + n \cdot 360° \ (n, \text{ any integer})$$

A graph of $y = \sin x$ and of $y = \frac{1}{2}$, the set of all points whose ordinates are $\frac{1}{2}$ (the dotted line) as shown in Figure 7.1 should illustrate the periodically recurring nature of the roots. Note that the abscissas of the points of

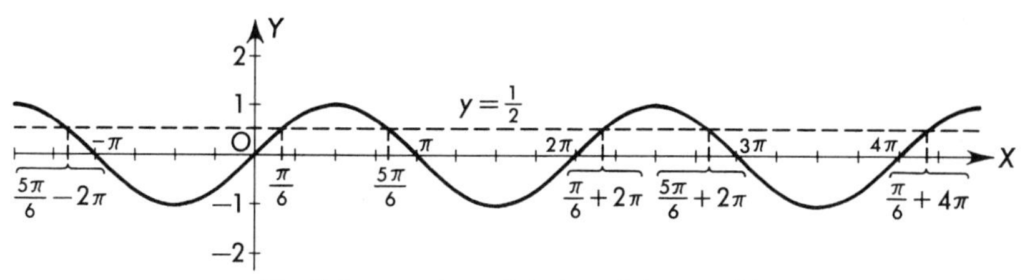


Fig. 7.1 $y = \sin x$. Solution of $\sin x = \frac{1}{2}$

intersection of the graph of $y = \sin x$ and the line $y = \frac{1}{2}$ form the solution set of the equation $\sin x = \frac{1}{2}$.

**EXAMPLE 2**

Approximate all values of $\theta$ in the interval $-360° \leqq 0 \leqq 360°$ that satisfy the equation

$$\sin \theta = -0.4536$$

**Solution:** From the table of natural functions (Table III in the back of the book),

$$\sin 26°\ 58' \approx 0.4536$$

Thus 26° 28′ is the reference angle (see Section 2.8). Since $\sin \theta$ is negative, $\theta$ is a third- or fourth-quadrant angle. Hence the desired values of $\theta$ in the

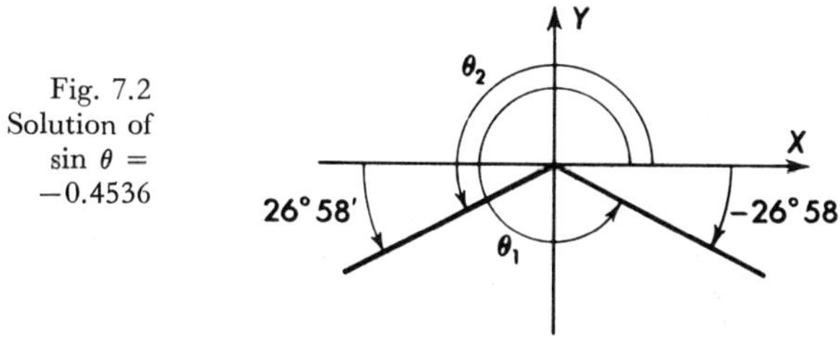


Fig. 7.2 Solution of $\sin \theta = -0.4536$

range $0° \leqq \theta \leqq 360°$ are (Figure 7.2)

$$\begin{aligned} \theta_1 &\approx 360° - 26°\ 58' \\ &= 333°\ 2' \end{aligned}$$

and

$$\theta_2 \approx 180° + 26°\ 58'$$
$$= 206°\ 58'$$

The other roots in the required interval are obtained by subtracting 360° from these. They are

$$\theta_3 = \theta_1 - 360°$$
$$\approx -\ 26°\ 58'$$

and

$$\theta_4 = \theta_2 - 360°$$
$$\approx -\ 152°\ 2'$$

**Conclusion**

There are four roots of $\sin\theta = -0.4536$ in the interval $-360° \leqq \theta \leqq 360°$ They are $-152°\ 2'$, $-26°\ 58'$, $206°\ 58'$, and $333°\ 2'$, approximately.

(*Note:* We could have determined $\theta_3$ and $\theta_4$ equally well by use of the reference angle and then found $\theta_1$ and $\theta_2$ by adding 360° to each of these.)

**EXAMPLE 3**

Find *all* roots of the equation

$$\tan x = \sqrt{3}$$

**Solution:** Since the period of tan $x$ is $\pi$, one needs only to find all solutions over an interval of one period. All other solutions will be found by adding multiples of $\pi$. Thus one must search for particular roots over the interval $0 \leqq x \leqq \pi$. By inspection, there is one such root, $x = \pi/3$. Therefore all roots are of the form

$$x = \frac{\pi}{3} + n\pi, \qquad n = 0, \pm 1, \pm 2, \cdots$$

The solution is shown graphically in Figure 7.3. Note how the points of

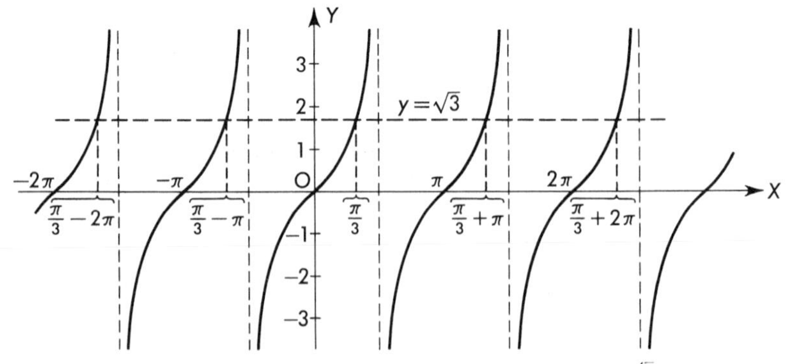


Fig. 7.3 Graph of $y = \tan x$. Solution of $\tan x = \sqrt{3}$

intersection of the curve $y = \tan x$ with the line $y = \sqrt{3}$ determine the values of $x$ which are the roots of $\tan x = \sqrt{3}$.

### EXAMPLE 4

Find all values of $x$ in the interval $0 \leqq x \leqq \pi$ that satisfy the equation

$$\cos 4x = \tfrac{1}{2}$$

**Solution:** Let $4x = \theta$. Then, when $x = 0$, $\theta = 0$, and, when $x = \pi$, $\theta = 4\pi$. Look for roots of $\cos \theta = \frac{1}{2}$ over the interval $0 \leqq \theta \leqq 4\pi$. By inspection, the required values of $\theta$ in the interval $0 \leqq \theta \leqq 2\pi$ are $\theta_1 = \pi/3$ and $\theta_2 = 5\pi/3$. The addition of $2\pi$ to each of these yields the other required values of $\theta$, $\theta_3 = \pi/3 + 2\pi = 7\pi/3$, and $\theta_4 = 5\pi/3 + 2\pi = 11\pi/3$. We now have:

$$\theta = \frac{\pi}{3}, \quad \frac{5\pi}{3}, \quad \frac{7\pi}{3}, \quad \text{and} \quad \frac{11\pi}{3}$$

But $4x = \theta$, so that $x = \frac{1}{4}\theta$. Hence the required values of $x$ are given by

$$x = \frac{\pi}{12}, \quad \frac{5\pi}{12}, \quad \frac{7\pi}{12}, \quad \text{and} \quad \frac{11\pi}{12}$$

Note that these values lie in the stated interval, $0 \leqq x \leqq \pi$.

## *EXERCISES*

**1.** Thinking of the points in common with the line $y = k$ and the curve $y = \sin x$ (Figure 7.1), determine all roots of each of the following equations.

(a) $\sin x = 1$ (b) $\sin x = 0$
(c) $\sin x = -1$ (d) $\sin x = 1.5$

**2.** Repeat Exercise 1 with *cosine* replacing *sine* throughout, and a sketch of $y = \cos x$.

**3.** Repeat Exercise 1 with *tangent* replacing *sine* throughout, using Figure 7.3.

**4.** Find all roots of each of the following equations for the interval indicated. If preferred, an equation involving $\cot x$, $\sec x$, or $\csc x$ may be replaced by one involving $\tan x$, $\cos x$, or $\sin x$, respectively. (How?)

(a) $\cos x = \dfrac{1}{\sqrt{2}}$, $\quad -\pi \leqq x \leqq \pi$

(b) $\tan x = -1$, $\quad 0 \leqq x \leqq 2\pi$

(c) $\sin x = \dfrac{1}{\sqrt{2}}$, $\quad 0° \leqq x \leqq 360°$

(**d**) $\sec x = \frac{2}{\sqrt{3}}$, $0° \leqq x \leqq 360°$
(**e**) $\sin x = 0.8910$, $-360° \leqq x \leqq 360°$
(**f**) $\cot x = -1$, $0 < x < 2\pi$
(**g**) $\cos x = 3/4$, $0 < x < 4\pi$
(**h**) $\sin x = 2$, $0 \leqq x \leqq 2\pi$
(**i**) $\tan x = \frac{1}{\sqrt{3}}$, all values of $x$

**5.** Find the roots of each of the following equations for the interval $0° \leqq x \leqq 360°$.

(**a**) $\sin 2x = 1/2$.
(**b**) $\cos \frac{x}{2} = 1/2$
(**c**) $\tan 3x = -1$
(**d**) $\cos 3x = 1$
(**e**) $\sin (x - 30°) = 1/2$
(**f**) $\tan (x + 45°) = 1$
(**g**) $\sin 3x = 1$
(**h**) $\sin \frac{2x}{3} = \frac{\sqrt{3}}{2}$

**6.** Find the four smallest positive roots of each of the following equations.

(**a**) $\cos \theta = 0$
(**b**) $\sin \varphi = 1$
(**c**) $\tan x = 1$
(**d**) $\cos \frac{t}{2} = -1$
(**e**) $\tan x = \sqrt{3}$
(**f**) $\sin 2x = 0$

---

## 7.3 INVERSE STATEMENT OF THE EQUATIONS $\sin x = k$, $\cos x = k$, AND SO FORTH

In Section 7.2 we discussed the solution of such equations as

$$\sin y = k$$

We found that for any permissible value of $k$ there is an infinite number of values of $y$ satisfying the equation. For instance, if $k = 1/2$, the equation is

$$\sin y = 1/2$$

and the roots are

$$\left.\begin{aligned} y &= \frac{\pi}{6} + n \cdot 2\pi \\ \text{and} \quad y &= \frac{5\pi}{6} + n \cdot 2\pi \end{aligned}\right\} \quad n, \text{ any integer}$$

There are many instances in which it is desirable to express $y$ as an element of the solution set of the equation $\sin y = k$ without assigning $k$ a particular value. To this end we say that

$y$ = *an angle (or a number) whose sine is* $k$

This *inverse* statement of the equation $\sin y = k$, expressing $y$ in terms of $k$, is of sufficient importance to give it the symbolic form

$$y = \text{arc} \sin k$$

or the optional form used by many writers,

$$y = \sin^{-1} k^*$$

Similarly, read *arc cos A* (or *cos*$^{-1}$ *A*) as *an angle whose cosine is A*, or as *inverse cosine A*. Likewise *arc tan x* (or *tan*$^{-1}$ *x*) is read *an angle whose tangent is x*, or *inverse tangent x*.

In general, then, the equations

$$x = \sin y$$

and

$$y = \text{arc} \sin x$$

are equivalent in the sense that any pair of values of $x$ and $y$ that make one of the equations a true statement also makes the other true. Other examples of equivalent inverse pairs of equations are: $y = 2x - 1$ and $x = (y + 1)/2$; $y = x^2$ and $x = \pm\sqrt{y}$; and $y = \log x$ and $x = 10^y$, always assuming that only permissible values of $x$ and $y$ are involved.

 **EXAMPLE 1**

Find all values of $\cos^{-1}(-\frac{1}{2})$.

**Solution:** If $\theta = \cos^{-1}(-\frac{1}{2})$, then, inversely, $\cos \theta = -\frac{1}{2}$. Hence $\theta$ terminates in quadrant II or quadrant III. By inspection, two values of $\theta$ are $2\pi/3$ and $4\pi/3$. (See Figure 7.4.) Thus, in general,

$$\cos^{-1}(-\tfrac{1}{2}) = \frac{2\pi}{3} + 2n\pi$$

and

$$= \frac{4\pi}{3} + 2n\pi$$

where $n$ is any integer.

---

* Note that the $-1$ is *not* an exponent. It is simply a part of the inverse symbol. If the $-1$ exponent is to be used, write $(\sin k)^{-1}$.

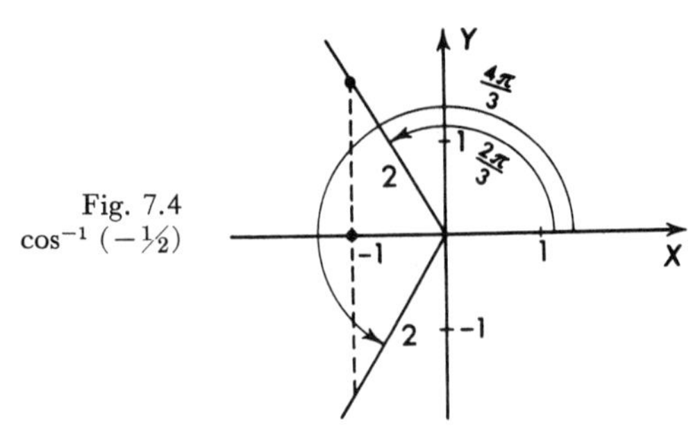


Fig. 7.4
$\cos^{-1}(-\frac{1}{2})$

[If preferred, or required, degree measure may be used, and the above result written

$$\left.\begin{aligned} \cos^{-1}(-\tfrac{1}{2}) &= 120° + n \cdot 360° \\ \text{and} \qquad &= 240° + n \cdot 360° \end{aligned}\right\} \quad n, \text{ any integer}]$$

**EXAMPLE 2**

Sketch the graph of

$$y = \text{arc sin } x$$

**Solution:** The equation

$$y = \text{arc sin } x$$

is equivalent to the equation

$$x = \sin y$$

the graph of which is shown in Figure 7.5. [Note that to each value of $x$ on

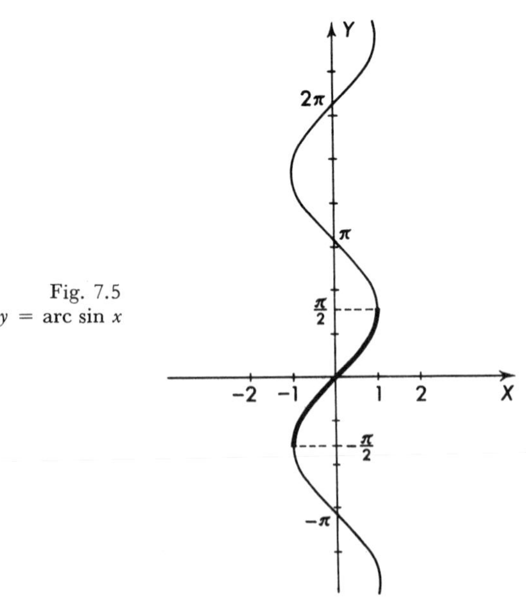


Fig. 7.5
$y = \text{arc sin } x$

the permissible interval ($-1 \leqq x \leqq 1$) there correspond an infinite number of values of $y$. The significance of the heavy portion of the curve will be explained in Section 7.4.]

**EXAMPLE 3**

Find $\theta = \text{arc sin}(-0.4536)$ if $-360° \leqq \theta \leqq 360°$.

**Solution:** Inversely stated,

$$\sin \theta = -0.4536$$

This was solved for $\theta$ in the required interval in Example 2, Section 7.2.

## 7.4 THE INVERSE TRIGONOMETRIC FUNCTIONS

Consider the sine function

$$\sin: A \rightarrow B$$

with domain $A$, the set of all angles, and $B$, the numbers of the interval $[-1, 1]$, recall that the function assigns to each angle $x_1$ exactly one number $y_1 = \sin x_1$, with $y_1$ in $B$. We say that the function *maps* $x_1$ into $y_1$ and that $y_1$ is the *image* of $x_1$. We say $x$ is the preimage of $y_1$. Question: Is there a function with domain $B$ and range $A$ that maps each $y_1$ of $B$ into its preimage in $A$? The answer is no, because each number from $-1$ to 1, inclusive is the sine of an infinite number of angles, as was seen in the previous section, while the definition of a function requires the assignment of exactly *one* element of $A$ to each $y_1$ of $B$. If, however, we restrict the domain of the sine function to the interval $[-\pi/2, \pi/2]$, there will be exactly one angle of this domain corresponding to each number of the range $[-1, 1]$. Let us denote this restricted sine function by use of a capital $S$ and define

$$\text{Sin}: A' \rightarrow B$$

to be the function with domain $A'$, the interval $[-\pi/2, \pi/2]$, and range $B$, the interval $[-1, 1]$, with $\text{Sin } x = \sin x$. This function has corresponding to it an inverse function which we call $\text{Sin}^{-1}$, or Arc sin (Note capital $S$ and $A$.), defined as follows:

$$\text{Sin}^{-1}: B \rightarrow A'$$

is the function that maps each element $x$ of $B$ into an element $y$ of $A'$ by the rule $y = \sin^{-1} x$. For instance, if $x = -\frac{1}{2}$, then $y = \sin^{-1} - \frac{1}{2} = -\pi/6$.

Corresponding to the Sine function and its inverse function, $\text{Sin}^{-1}$ (or Arc sin), we have the Cosine function and its inverse, and the Tangent function and its inverse. Thus:

$$\text{Cos}: A'' \rightarrow B, \text{ with Cos } x = \cos x$$

and with domain $A''$, the set of angles with measures in the interval $[0, \pi]$, and range $B$ the number interval $[-1, 1]$, and

$$\text{Cos}^{-1}: B \rightarrow A'', \text{ with } \text{Cos}^{-1}\, x = \cos^{-1} x$$

Also the Tangent function with domain the set of angles of measures in the interval $(-\pi/2, \pi/2)$. (Note the exclusion of $-\pi/2$ and $\pi/2$) and range the set of all numbers. As with the others, $\text{Tan}\, x = \tan x$. This has the inverse Tangent function $\text{Tan}^{-1}$ that interchanges domain and range with Tan and is such that $\text{Tan}^{-1}\, x = \tan^{-1} x$.

Restricted-range Cot, Sec, and Csc functions and their inverses can be devised by assigning suitable domains, but these will not be treated in this book. In most applications the above three functions and their inverses are sufficient.

We note that the symbols $\text{Sin}^{-1}$ and Arc sin are interchangeable, as are $\text{Cos}^{-1}$ and Arc cos, and likewise $\text{Tan}^{-1}$ and Arc tan. While the sets of angles

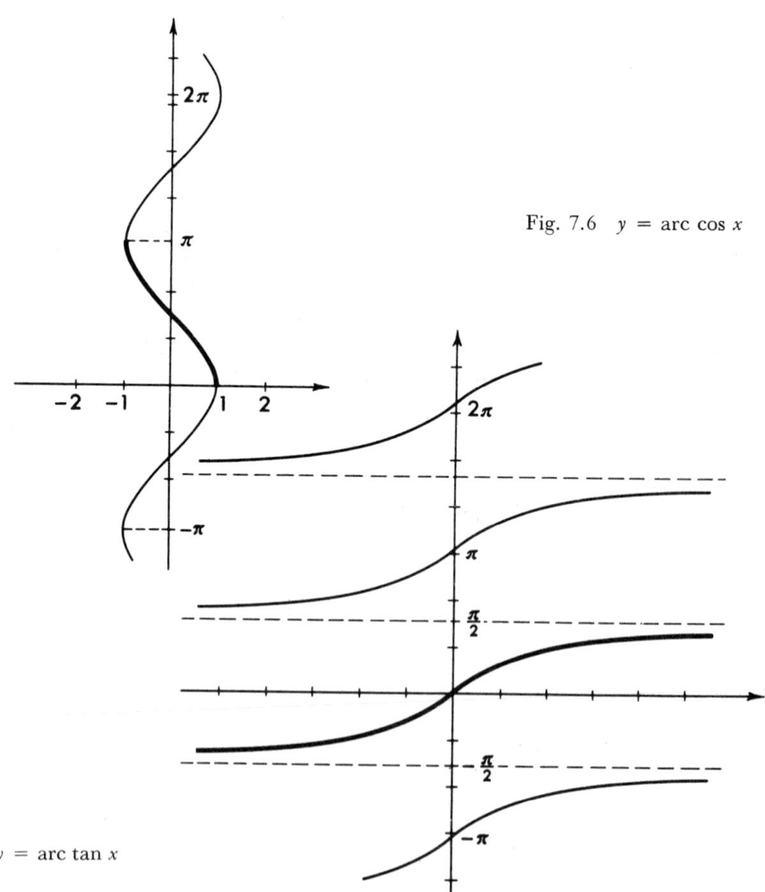


Fig. 7.6 $y = \text{arc cos } x$

Fig. 7.7 $y = \text{arc tan } x$

involved in each of these functions may be given in degree measure, it is best to use radian measure because that is the measure used almost exclusively in calculus. Furthermore Sin, Cos, and Tan may be functions of *numbers* instead of angles if the nature of the situation makes this appropriate.

The graph of $y = \text{Sin}^{-1} x$ consists of the heavy part of the curve of Figure 8.5. Note that the domain is the number interval $[-1, 1]$ and the range is the set of angles (or numbers) from $-\pi/2$ to $\pi/2$, inclusive. The graphs of $y = \text{Cos}^{-1} x$ and $y = \text{Tan}^{-1} x$ are the heavy portions of the graphs of $y = \cos^{-1} x$ and $y = \tan^{-1} x$ shown in Figures 7.6 and 7.7.

The extent of the domain and range for each of the three inverse functions is shown as follows:

| Function | Domain | Range |
|---|---|---|
| Arc sin | $-1 \leqq x \leqq 1$ | $-\pi/2 \leqq \text{Arc sin } x \leqq \pi/2$ |
| Arc cos | $-1 \leqq x \leqq 1$ | $0 \leqq \text{Arc cos } x \leqq \pi$ |
| Arc tan | $x$, any number | $-\pi/2 < \text{Arc tan } x < \pi/2$ |

We observe that the expression Arc sin $x$ may be interpreted to mean "the angle with measure of smallest absolute value whose sine is $x$." The same is true of Arc cos $x$ and Arc tan $x$, except in the case of Arc cos $x$, the angle is to be that of smallest *positive* measure with cosine equal to $x$.

### EXAMPLE 1

Find Arc cos $(-\frac{1}{2})$.

**Solution:** The range of the Arc cos function is from 0 to $\pi$, inclusive. Hence we look for a root of the equation $\cos x = -\frac{1}{2}$ in that range. The answer is $x = 2\pi/3$. Hence Arc cos $(-\frac{1}{2}) = 2\pi/3$.

### EXAMPLE 2

Evaluate cos [Arc tan $(-\frac{4}{3})$].

**Solution:** Sketch Arc tan $(-\frac{4}{3})$ in standard position as shown in Figure 7.8. Let $P(x, y) = (3, -4)$, and

$$r = \sqrt{3^2 + 4^2} = 5$$

Hence

$$\cos [\text{Arc tan } (-\tfrac{4}{3})] = \frac{x}{r} = \tfrac{3}{5}$$

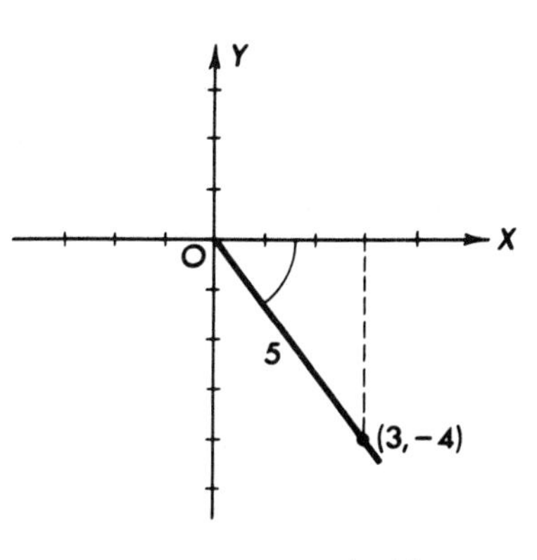


Fig. 7.8 Arc tan $(-\frac{4}{3})$

### EXAMPLE 3

Simplify cos [Arc sin $(-\frac{3}{5})$ − Arc cos $\frac{3}{5}$].

**Solution:** Let $\theta$ = Arc sin $(-\frac{3}{5})$ and let $\varphi$ = Arc cos $(\frac{3}{5})$. Then

$$\begin{aligned}\cos\,[\text{Arc sin}\,(-\tfrac{3}{5}) - \text{Arc cos}\,\tfrac{3}{5}] &= \cos\,(\theta - \varphi)\\ &= \cos\theta\cos\varphi + \sin\theta\sin\varphi\\ &= (\tfrac{4}{5})(\tfrac{3}{5}) + (-\tfrac{3}{5})(\tfrac{4}{5})\\ &= 0\end{aligned}$$

### EXAMPLE 4

Find a *single* general formula for all roots of the equation below. Use radian measure.

$$\sin x = k, \qquad -1 \leqq k \leqq 1$$

**Solution:** One solution is $x$ = Arc sin $k$. Since $\sin(\pi - \theta) = \sin\theta$, another solution within one cycle is $\pi$ − Arc sin $k$. Hence we have as roots of $\sin x = k$ the following:

$$\left.\begin{aligned} x &= \text{Arc sin}\,k + n\cdot 2\pi,\\ \text{and}\qquad &\pi - \text{Arc sin}\,k + n\cdot 2\pi\end{aligned}\right\}\quad n,\text{ any integer}$$

These may be rewritten

$$\left.\begin{aligned} x &= 2n\pi + \text{Arc sin}\,k,\\ \text{and}\qquad &(2n+1)\pi - \text{Arc sin}\,k\end{aligned}\right\}\quad n,\text{ any integer}$$

The single formula

$$x = m\pi + (-1)^m\,\text{Arc sin}\,k, \quad m,\text{ any integer}$$

is equivalent to the two results above, even values of $m$ giving the first line, and odd values of $m$ giving the second. Hence this is the required formula.

## *EXERCISES*

Find the value of each of the following, if it exists.

1. Arc cos $\frac{1}{2}$
2. $\text{Sin}^{-1}\dfrac{\sqrt{3}}{2}$
3. $\text{Tan}^{-1}(-1)$
4. Arc cos 0
5. Arc tan $(-1/\sqrt{3})$
6. $\text{Cos}^{-1}\,2$
7. $\text{Sin}^{-1}\,0 + \text{Tan}^{-1}\sqrt{3}$
8. Arc cos 1 + Arc tan 0
9. $\text{Cos}^{-1}(\sin 90°)$
10. $\text{Sin}^{-1}(\sin 270°)$
11. Arc tan $(\tan[-\pi/4])$
12. Arc sin $(\cos 50°)$
13. Arc cos $(\sin[-100°])$
14. $\text{Sin}^{-1}(\tan 0.23)$
15. sin (2 Arc sin $\frac{1}{2}$)
16. sin (2 Arc cos $\frac{4}{5}$)

**17.** $\tan \frac{1}{2}$ (Arc sin 5/13)
**18.** $\cos \frac{1}{2}$ (Arc tan 24/25)
**19.** $\cos (2 \operatorname{Tan}^{-1} [-1])$
**20.** Arc sin $(-0.6691)$
**21.** $\sin (\operatorname{Sin}^{-1} \frac{1}{2} + \operatorname{Cos}^{-1} \frac{3}{5})$
**22.** cos (Arc sin $y$) (check for positive and negative values of $y$)
**23.** $\tan (\operatorname{Sin}^{-1} y)$
**24.** tan (2 Arc tan 2)
**25.** cos (Arc sin $x$ + Arc sin $y$)
**26.** $\tan (\operatorname{Tan}^{-1} x + \operatorname{Tan}^{-1} 2y)$
**27.** Arc cos {sin [Arc tan $(-1)$]}
**28.** $\cos^2 (\frac{1}{2} \operatorname{Cos}^{-1} 2x)$

**29.** Solve the following for $x$ in terms of $y$.

(**a**) $y = 2$ arc sin $x$
(**b**) $y = \frac{1}{2}$ arc cos $\frac{1}{2}x$
(**c**) $y = 2 \sin^{-1} x/3$
(**d**) $y = \frac{2}{3}$ arc cot $\frac{1}{4}x$

**30.** Prove that each of the following equations is a true statement.

(**a**) 2 Arc tan $\frac{2}{3}$ = Arc tan 12/5. *Hint:* Each side represents an angle. Show that these angles have equal tangents. However, you must also argue that the angles have the same measure.
(**b**) Arc sin $\frac{4}{5}$ + Arc sin 8/17 + Arc sin 13/85 = $\pi/2$. (See Exercise 18 of Section 6.9.)
(**c**) Arc sin 11/16 = 3 Arc sin $\frac{1}{4}$.

**31.** Prove that

$$\cos \left\{ \operatorname{Tan}^{-1} \left[ \sin \left( \operatorname{Tan}^{-1} \frac{1}{x} \right) \right] \right\} = \sqrt{\frac{x^2 + 1}{x^2 + 2}}$$

**32.** Explain why the range for Arc sin $x$, $-\pi/2 \leqq$ Arc sin $x \leqq \pi/2$, should include the end values $-\pi/2$ and $\pi/2$, while the range for Arc tan $x$, $-\pi/2 <$ Arc tan $x < \pi/2$, excludes the end values.

**33.** Sketch $y = \sin x$ and $y = \sin^{-1} x$ on the same axes. Note that, if the paper is folded along the line bisecting the first quadrant ($\theta = 45°$), the two curves coincide. Why is this? Is the same true of the graphs of $y = \cos x$ and $y = \cos^{-1} x$? of $y = \tan x$ and $y = \tan^{-1} x$?

**34.** Change the formula for the general solution of $\sin x = k$ given in Example 4 above to degree measure. Find the solutions of

$$\sin x = \tfrac{1}{2}$$

corresponding to $m = 0$, $m = 1$, $m = 6$.

**35.** Using the fact that one solution of the equation $\cos x = k$ is $x =$ Arc cos $k$, show that a single formula for all solutions of the equation

$$\cos x = k, \quad -1 \leqq k \leqq 1$$

is

$$x = 2m\pi \pm \text{Arc cos } k, \quad m, \text{ any integer}$$

[*Hint:* Remember that $\cos(-\theta) = \cos \theta$, so that, if $x = \theta$ is a solution of $\cos x = k$, then so also is $x = -\theta$.]

**36.** Use the formula of Exercise 35 to find the four smallest positive roots of

$$\cos x = \tfrac{1}{2}$$

**37.** Remembering that the period of the function $y = \tan x$ is $\pi$, find a general formula for the solutions of the equation

$$\tan x = k$$

Use this formula to find the four smallest positive roots of

$$\tan x = -1$$

**38.** Show that the equations below are identities (mentioning the values of $x$ for which at least one member is meaningless). (*Note:* These identities are the basis for the fact that the inverse cotangent, inverse secant, and inverse cosecant are infrequently used.)

(**a**) $\operatorname{arc}\cot x = \operatorname{arc}\tan \dfrac{1}{x}$

(**b**) $\operatorname{arc}\sec x = \operatorname{arc}\cos \dfrac{1}{x}$

(**c**) $\operatorname{arc}\csc x = \operatorname{arc}\sin \dfrac{1}{x}$

---

## 7.5 SOLUTION OF OTHER FORMS OF TRIGONOMETRIC EQUATIONS

In general, the solutions of trigonometric equations of more complicated forms are effected by reducing the equations to systems of equations of the form treated in Section 7.2. No one method of reduction covers all types, and frequently the reduction is a matter of ingenuity. We discuss a few principles of equation solving and then illustrate the methods of solving some of the frequently encountered types.

It must be emphasized that, when an algebraic operation is performed on both members of an equation, the result is a *different* equation. It may have exactly the same roots as the given equation, in which case the equations are said to be *equivalent;* or it may have roots that are *not* roots of the original equation. The important thing is that, if we avoid dividing both members of the given equation by an expression containing the unknown, the roots of the given equation are *among* those of the derived equation. Checking in the original equation may be necessary to determine which roots of the derived equation are roots of the given equation, and *such checking is part of the solution process.* An equation that is derived from a given equation by the addition of a term to both members (or subtraction of a term from both members) is equivalent to the given equation. Checking is not a necessary part of the solution in this case. The same can be said if both members of an equation are multiplied or divided by a nonzero *constant*. All other operations should be used with caution and answers should be checked.

Before presenting some illustrative examples, we offer the following suggestions which may be found useful in reducing trigonometric equations to the elementary types:

(1) Try to express all functions involved in terms of a single function of some angle.

(2) Collect all terms on one side and equate to zero.

(3) Factor, if possible, or use the quadratic formula* if the equation is of second degree in the unknown function.

(4) Using the principle that $A \cdot B = 0$ if, and only if, $A = 0$ or $B = 0$, set each factor in (3) equal to zero, thus obtaining a simpler set of equations.

### EXAMPLE 1

*Reducible by Algebraic Factoring*. Solve the equation

$$2 \sin x \cos x + 2 \sin x - \cos x - 1 = 0$$

for roots in the interval $0 \leqq x \leqq 3\pi$.

**Solution:** Removal of the factor $2 \sin x$ from the first two terms and the factor $-1$ from the last two gives

$$2 \sin x (\cos x + 1) - 1 (\cos x + 1) = 0$$

Removing the common factor $(\cos x + 1)$ results in

$$(\cos x + 1)(2 \sin x - 1) = 0$$

from which

$$\cos x + 1 = 0 \quad \text{or} \quad 2 \sin x - 1 = 0$$

The first of these can be written

$$\cos x = -1$$

By inspection, the roots of this equation are

$$x = \pi + n \cdot 2\pi$$

* The two solutions of the equation $ax^2 + bx + c = 0$ are

$$x_1 = \frac{-b + \sqrt{b^2 - 4ac}}{2a} \quad \text{and} \quad x_2 = \frac{-b - \sqrt{b^2 - 4ac}}{2a}$$

and hence roots in the specified interval are $x = \pi$ and $x = 3\pi$. The second equation can be written

$$\sin x = \tfrac{1}{2}$$

This equation was solved as an illustration in Section 7.2. The roots in the required range are $x = \frac{\pi}{6}, \frac{5\pi}{6}, \frac{13\pi}{6}, \frac{17\pi}{6}$. Finally, all roots of the given equation in the range $0 \leqq x \leqq 3\pi$ are $x = \frac{\pi}{6}, \frac{5\pi}{6}, \pi, \frac{13\pi}{6}, \frac{17\pi}{6}, 3\pi$. The checking of these roots, while desirable, is not an essential part of the solution.

### EXAMPLE 2

*Reducible with the Aid of Trigonometric Identities.* Solve for values of $x$ between 0° and 360°,

$$3 \cos x + \sin x = 1$$

**Solution:** Subtracting $3 \cos x$ from both members and squaring the resulting members give

$$\sin^2 x = (1 - 3 \cos x)^2$$

But $\sin^2 x = 1 - \cos^2 x$, so that

$$1 - \cos^2 x = 1 - 6 \cos x + 9 \cos^2 x$$

which reduces to

$$5 \cos^2 x - 3 \cos x = 0$$

or

$$\cos x\, (5 \cos x - 3) = 0$$

Hence

$$\cos x = 0 \quad \text{or} \quad \cos x = \tfrac{3}{5}$$

The solutions of $\cos x = 0$ in the required interval are $x = 90°$ and $x = 270°$. The solutions of $\cos x = \frac{3}{5}$ are $x = 58° \ 8'$ and $x = 306° \ 52'$. These must be checked. (Why?) Substitution of each of these four possible roots in the original equation reveals that only $x = 90°$ and $x = 306° \ 52'$ are solutions.

### EXAMPLE 3

*Reducible by the Factor Formulas.* Find all roots of the following equation which are in the interval $-\pi/2 \leqq x \leqq \pi$:

$$\sin 5x + \sin 3x + \cos x = 0$$

**Solution:** By applying formula (6.26) to $\sin 5x + \sin 3x$ (or by using $5x = 4x + x$ and $3x = 4x - x$ and applying the sine addition formulas), the given equation may be reduced to the equivalent equation

$$2 \sin 4x \cos x + \cos x = 0$$

which in factored form is

$$\cos x\,(2 \sin 4x + 1) = 0$$

Hence

$$\cos x = 0 \quad \text{or} \quad \sin 4x = -\tfrac{1}{2}$$

The solutions of the equation $\cos x = 0$ in the given interval are

$$x = -\frac{\pi}{2} \quad \text{and} \quad x = \frac{\pi}{2}$$

To find the solutions of the equation $\sin 4x = -\frac{1}{2}$ on the interval $-\pi \leqq x \leqq \pi$, first notice that, when $x = \pm\pi$, $4x = \pm 4\pi$. Thus look for values of $4x$ such that $-4\pi \leqq 4x \leqq 4\pi$. (See Example 3, Section 7.2.) By inspection, two values of $4x$ that satisfy the equation $\sin 4x = -\frac{1}{2}$ are

$$4x = \frac{7\pi}{6} \quad \text{and} \quad 4x = \frac{11\pi}{6}$$

so that general values are

$$4x = \frac{7\pi}{6} + 2n\pi \quad \text{and} \quad 4x = \frac{11\pi}{6} + 2n\pi$$
$$n = 0, \pm 1, \pm 2, \ldots$$

Of these values of $4x$, the following lie in the required interval:

$$\text{with } n = -2, \qquad 4x = -\frac{17\pi}{6} \quad \text{and} \quad -\frac{13\pi}{6}$$

$$\text{with } n = -1, \qquad 4x = -\frac{5\pi}{6} \quad \text{and} \quad -\frac{\pi}{6}$$

$$\text{with } n = 1, \qquad 4x = \frac{19\pi}{6} \quad \text{and} \quad \frac{23\pi}{6}$$

To these six values of $4x$ correspond the six values of $x$:

$$x = -\frac{17\pi}{24}, \; -\frac{13\pi}{24}, \; -\frac{5\pi}{24}, \; -\frac{\pi}{24}, \; \frac{19\pi}{24}, \text{ and } \frac{23\pi}{24}$$

These, and the values $x = -\pi/2$ and $x = \pi/2$, found above, form the entire set of solutions of the original equation over the interval $-\pi \leqq x \leqq \pi$. Checking is not essential.

## *EXERCISES*

In Exercises **1** through **12**, find the roots from 0 to $2\pi$, inclusive. Find also the general solution sets.

**1.** $2 \sin^2 x - \sin x - 1 = 0$
**2.** $4 \cos^2 \theta - 4 \sin^2 \theta + 1 = 0$
**3.** $\cot \theta = \tan \theta$
**4.** $\cos 2x = \cos x$
**5.** $\sin 2y = 2 \cos y$
**6.** $\sin 3y = -1/2$
**7.** $\sin x + \sin 2x + \sin 3x = 0$
**8.** $\sin 4x + \cos 2x = 0$
**9.** $\cos x - 1 = \sin \frac{x}{2}$
**10.** $4 \cos^2 x - 3 = 0$
**11.** $3 \tan x = 2 \cos x$
**12.** $\cos x - \cos 2x = \sin x$

In Exercises **13** through **20**, find all roots in the interval $0 \leqq x \leqq 360°$

**13.** $\cos x + \cos 3x + \cos 5x = 0$
**14.** $3 \cos x + 4 \sin x = 3$
**15.** $\sin 3x \cos x + \cos 3x \sin x = 1/2$
**16.** $\cos x - \sin x - 1 = 0$
**17.** $\sin^2 \frac{x}{2} + \cos x + \cos^2 \frac{x}{2} = 2$
**18.** $3 \sin x \tan x - 9 \sin x - \tan x + 3 = 0$
**19.** $2 \sin x \cos x + \cos x - 2 \sin x = 1$
**20.** $3 \tan x - \cot x = 2$
**21.** Solve for $x$: $\text{Arc} \sin x + \text{Arc} \sin 2x = \pi/2$
*Hint:* Take the cosine of each side of the equation.
**22.** Solve for $x$: $\text{Sin}^{-1} (1 - x^2) = 2 \text{ Sin}^{-1} 3/5$
**23.** Solve for $y$: $\text{Arc} \tan 2y + \text{Arc} \tan 3y = \pi/4$
**24.** Solve for $t$: $\text{Arc} \sin 2t - \text{Arc} \cos t = \pi/6$

## *MISCELLANEOUS EXERCISES/CHAPTER 7*

In Exercises **1** through **12**, determine which of the equations are identities. Solve those that are not for roots between 0 and $2\pi$, inclusive.

**1.** $(\sin x + \cos x)^2 = \sin 2x + 1$
**2.** $\sin x \cos x - \cot x = 0$
**3.** $\tan 2\theta + 2 \sin \theta \cos \theta = 0$

**4.** $\tan \frac{y}{2} + \cot \frac{y}{2} = 2 \csc y$

**5.** $\sin 3\theta + \sin^2 \theta = 0$

**6.** $\tan \frac{x}{2} + \cos x = \csc x$

**7.** $\frac{1 - \cos 2\theta}{1 + \cos 2\theta} = \tan^2 \theta$

**8.** $\sin z + 2 \cos z = 2$

**9.** $\sin 2x - \cos 2x = \sin x - \cos x$

**10.** $\sin 2\theta = \cot \theta\,(1 - \cos 2\theta)$

**11.** $\cos^4 \theta - \sin^4 \theta = \cos 2\theta$

**12.** $\tan (x + \pi/4) = (1 + \sin 2x)/\cos 2x$

**13.** Given that $\theta$ is a positive acute angle, express each of the following in terms of $\theta$ (without mention of the trigonometric functions of $\theta$ in the result).

(**a**) $\text{Cos}^{-1} (-\sin \theta)$
(**b**) $\text{Arc sin} [\sin (\theta + \pi/2)]$
(**c**) $\text{Arc sin} [\cos (\pi - \theta)]$
(**d**) $\text{Sin}^{-1} [\cos (-\theta)]$

**14.** Solve each of the following for $x$.

(**a**) $\text{Cos}^{-1} x + \text{Cos}^{-1} 2x = \pi/2$
(**b**) $\text{Arc sin } x + \text{Arc sin } 2x = \pi/3$
(**c**) $\text{Sin}^{-1} \frac{1}{3} + \text{Tan}^{-1} \frac{1}{2} = \text{Sin}^{-1} x$

**15.** Find the value of:

(**a**) $\sin \frac{1}{2} (\text{Arc tan } \frac{3}{4})$
(**b**) $\cos [2\, \text{Tan}^{-1} 5/12]$
(**c**) $\sec [\frac{1}{2} \text{Arc tan } 7/24]$
(**d**) $\tan (2\, \text{Sin}^{-1} \frac{1}{2})$

**16.** Show that each of the following is a true statement.

(**a**) $\text{Arc sin } \frac{1}{2} + 2\, \text{Arc cos } \sqrt{3}/2 = \pi/2$
(**b**) $2\, \text{Arc tan } 1 - \frac{1}{2} \text{Arc sec } 2 = \pi/3$
(**c**) $\text{Sin}^{-1} 1/\sqrt{5} + \text{Sin}^{-1} 2/\sqrt{5} = \pi/2$
(**d**) $\text{Tan}^{-1} \frac{1}{2} + \text{Tan}^{-1} \frac{1}{5} + \text{Tan}^{-1} \frac{1}{8} = \pi/4$
(**e**) $\text{Arc tan } \frac{3}{2} - \text{Arc tan } \frac{1}{5} = \pi/4$

(It is interesting to note that these relations are used with the aid of series in calculus to evaluate $\pi$ to the desired degree of accuracy.)

**17.** Given the pair of equations

$$\rho \sin \theta = 3$$

and

$$\rho \cos \theta = 4$$

find values of $\rho$ and $\theta$ which are a simultaneous solution of the equations and are such that $\rho > 0$ and $0 \leqq \theta \leqq 2\pi$. (*Hint:* Square both members of each equation and add.)

**18.** Find algebraically the simultaneous solutions of the pair of equations

$$\rho \sin \theta = 1$$

and

$$\rho = 1 + 2 \sin \theta$$

for $\rho > 0$ and $0 \leqq \theta \leqq 2\pi$. Check by sketching in polar coordinates and observing the points of intersection.

**19.** Find the simultaneous solutions of the pair of equations

$$\rho = 2 \sin \theta$$

and

$$\rho = 2 \cos \theta$$

for $\rho > 0$ and $0° \leqq \theta \leqq 360°$. Check by sketching in polar-coordinates and examining the points of intersection of the curves. Note that these curves cross at the origin. Can you explain why this common point is not discovered by your simultaneous solution?

**20.** In calculus it is helpful to be able to simplify certain expressions involving inverse trigonometric functions.

(**a**) Verify that $\text{Arc tan} \dfrac{x}{\sqrt{1 - x^2}} = \text{Arc sin}\, x$ for $-1 < x < 1$.

(**b**) Verify that $\text{Arc cos} \dfrac{1}{\sqrt{x^2 + 1}} = \text{Arc tan}\, |x|$ for all values of $x$.

(**c**) Simplify $\text{Arc cos} \dfrac{\sqrt{x^2 + 2x}}{x + 1}$ if $x > 0$

**21.** Making use of the idea of Example 4, Section 7.4, verify that:

(**a**) $\text{arc sin}\, k = n\pi + (-1)^n\, \text{Arc sin}\, k$, for $-1 \leqq k \leqq 1$
(**b**) $\text{arc cos}\, k = 2n\pi \pm \text{Arc cos}\, k$, for $-1 \leqq k \leqq 1$
(**c**) $\text{arc tan}\, k = n\pi + \text{Arc tan}\, k$, for every number $k$

**22.** Use the formulas of Exercise 21 to give the general solution sets (in formula form) for the following equations.

(**a**) $\sin \theta \cos \theta = 0$
(**b**) $\sin \theta + \sin \theta \cos \theta = 0$
(**c**) $4 \sin^2 \theta = 1$
(**d**) $\cos 2\theta = \dfrac{1}{\sqrt{2}}$

# CHAPTER 8. THE GENERAL TRIANGLE

## 8.1 SOLUTION OF THE GENERAL TRIANGLE

A triangle is determined if a side and two other parts (angles or sides) are known.* The unknown parts may be found with the aid of the formulas developed in this chapter. Occasions often arise in applied mathematics in which a statement of the relationship among parts of a triangle is of more interest than a numerical solution for the missing parts. The formulas of this chapter, particularly the *law of sines* and the *law of cosines*, provide the means for this.

Although right triangles may be solved by the methods of this chapter, it should be remembered that they are more easily solved by the direct application of the definitions of the trigonometric functions as presented in Chapter 5.

We continue to use the notation of Chapter 5 for triangle $ABC$: $\alpha$, $\beta$, $\gamma$ for the angles with vertices at $A$, $B$, $C$, respectively, and $a$, $b$, $c$ for the corresponding opposite sides. Letters $A$, $B$, $C$ will also represent angles $\alpha$, $\beta$, $\gamma$, respectively, if no confusion results.

## 8.2 THE LAW OF SINES

In any triangle $ABC$,

$$\frac{a}{\sin \alpha} = \frac{b}{\sin \beta} = \frac{c}{\sin \gamma} \tag{8.1}$$

This relation of sides to sines of angles is known as the law of sines. In words: *In any triangle the sides are proportional to the sines of the opposite angles.*

* In case the known parts are two sides and an angle opposite one of them, the triangle, if determined, may not be unique. This case will be discussed in Section 8.2 as the *ambiguous case*.

**PROOF**

Apply a coordinate system to the triangle so that an angle, say angle $\alpha$, is in standard position. [Figure 8.1 illustrates this, with $\alpha$ an acute angle

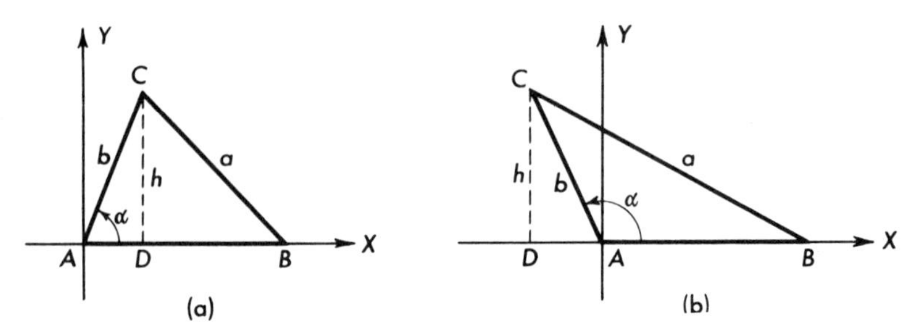


Fig. 8.1 Triangle $ABC$ with $\alpha$ in standard position

in (a), and $\alpha$ an obtuse angle in (b). The statements to follow apply to either figure.] Then $h$, the length of the altitude of the triangle from $C$, is also the abscissa ($y$ coordinate) of $C$. Hence, from the definition of sin $\alpha$,

$$h = b \sin \alpha$$

Also, from right triangle $CDB$,

$$h = a \sin \beta$$

Thus

$$b \sin \alpha = a \sin \beta$$

so that

$$\frac{a}{\sin \alpha} = \frac{b}{\sin \beta}$$

Similarly, the placing of $\beta$ in standard position results in

$$\frac{b}{\sin \beta} = \frac{c}{\sin \gamma}$$

which completes the proof.

The equation form of the law of sines (8.1) is actually a set of three equations each of which states the proportionality of two sides of a triangle to the sines of the angles opposite them. A study of these equations shows that triangles with two angles and a side known or with two sides and an angle opposite one of them known can be solved by use of the law of sines. The following examples illustrate these cases.

**EXAMPLE 1**

A side of a parallelogram 598 ft long makes angles of 27° and 42° with the diagonals. Find the lengths of the diagonals.

**Solution:** Since the diagonals of the parallelogram bisect each other, $OA$ and $OB$ (Fig. 8.2) can be calculated and respectively doubled. To apply

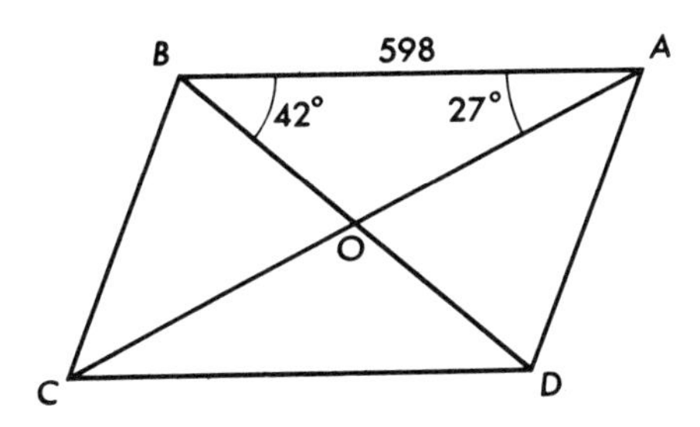


Fig. 8.2 Example 1

the law of sines in solving triangle $AOB$, the angle opposite $AB$ is needed. It may be computed by subtracting the sum of the two known angles from 180° and is found to be 111°. Then, from the law of sines,

$$\frac{OB}{\sin 27^\circ} = \frac{598}{\sin 111^\circ}$$

But $\sin 111^\circ = \sin 69^\circ$. Hence

$$\frac{OB}{\sin 27^\circ} = \frac{598}{\sin 69^\circ}$$

or

$$\begin{aligned} \log OB &= \log \sin 27^\circ + \log 598 - \log \sin 69^\circ \\ &\approx (9.6570 - 10) + 2.7767 - (9.9702 - 10) \\ &= 2.4635 \end{aligned}$$

Therefore

$$OB \approx 291$$

and

$$BD \approx 582$$

Similarly, from the equation

$$\frac{OA}{\sin 42^\circ} = \frac{598}{\sin 111^\circ}$$

it follows that

$$AC \approx 857$$

**EXAMPLE 2**

A farmer owns a tract of land bounded on two sides by straight highways intersecting at an angle of 38°. He decides to fence in a triangular field measured off as follows: From $A$, the corner of his property at the highway intersection, he will measure along one highway a distance 1000 ft to point $B$. From $B$ he will run a straight fence 800 ft long to point $C$ on the second highway. The final side of the triangular field will be side $CA$. How long is $CA$?

**Solution:** We are given triangle $ABC$ with $\alpha = 38°$, $c = 1000$, and $a = 800$; $a$ and $b$ are assumed to be accurate to the nearest foot. We attempt to find $\gamma$, from which the value of $\beta$ follows (since $\alpha + \beta + \gamma = 180°$). Side $b$ can then be computed.

From the law of sines,

$$\frac{\sin \gamma}{c} = \frac{\sin \alpha}{a}$$

Hence

$$\frac{\sin \gamma}{1000} = \frac{\sin 38°}{800}$$

from which

$$\sin \gamma \approx \frac{(1000)(0.6157)}{800}$$
$$\approx 0.7696$$

Remembering that $\sin \theta = \sin (180° - \theta)$, we observe that there are two angles each of whose sine is 0.7696, namely $50°\ 20'$ and $180° - 50°\ 20' = 129°\ 40'$, approximately. Naming these possibilities $\gamma$ and $\gamma_1$, we have

$$\gamma \approx 50°\ 20'$$

and

$$\gamma_1 \approx 129°\ 40'$$

These results are shown in Figure 8.3. Thus there are two equally possible triangles, triangle $ABC$ and triangle $ABC_1$.

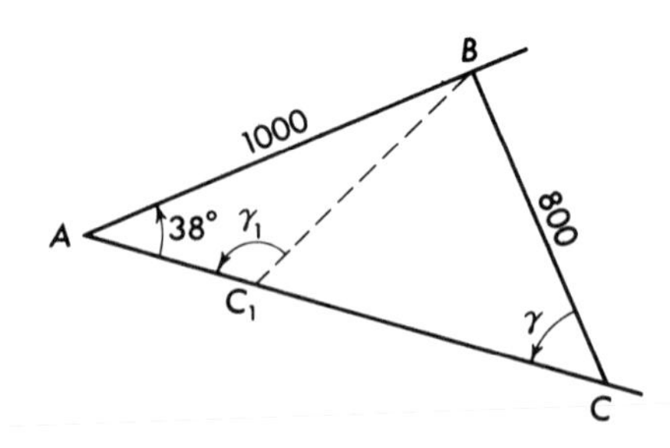


Fig. 8.3 Example 2

In triangle $ABC$,

$$\beta = 180° - (\alpha + \gamma)$$
$$\approx 91°\ 40'$$

To find $b$, we again use the law of sines. Thus

$$\frac{b}{\sin \beta} = \frac{a}{\sin \alpha}$$

from which

$$\frac{b}{\sin 91^\circ 40'} = \frac{800}{\sin 38^\circ}$$

Again noting that $\sin 91^\circ 40' = \sin (180^\circ - 91^\circ 40') = \sin 88^\circ 20'$, we obtain

$$\begin{aligned} b &\approx \frac{(0.9999)(800)}{0.6157} \\ &= 1300 \text{ ft, to the nearest 10 ft} \end{aligned}$$

On the other hand, in triangle $ABC_1$,

$$\begin{aligned} \beta_1 &= 180^\circ - (\alpha + \gamma_1) \\ &\approx 12^\circ 20' \end{aligned}$$

Now

$$\frac{b_1}{\sin \beta_1} = \frac{a}{\sin \alpha}$$

and thus

$$\frac{b_1}{\sin 12^\circ 20'} = \frac{800}{\sin 38^\circ}$$

Hence

$$\begin{aligned} b_1 &\approx \frac{(0.2136)(800)}{0.6157} \\ &\approx 277 \text{ ft} \end{aligned}$$

**CONCLUSION**

The farmer will have to make a choice between two possibilities, both equally possible from his original description.

[*Note:* The case of a triangle in which two sides and the angle opposite one of them are given is called the *ambiguous case* because various possibilities exist. The equation

$$\sin \gamma = \frac{c \sin \alpha}{a}$$

arises in applying the law of sines to find $\gamma$ when $c$, $a$, and $\alpha$ are known. This equation in $\gamma$ will have no solution, one solution, or two solutions, according as $(c \sin \alpha)/a$ is greater than 1, equal to 1, or less than 1. Corresponding to these possibilities, we have no triangle, one triangle, or *possibly* two triangles (two triangles if the $\alpha + \beta + \gamma = 180^\circ$ law permits, as in Example 2 above; otherwise, a single triangle, as in Example 3 below).]

**EXAMPLE 3**

Solve triangle $ABC$ with $a \approx 12$, $b \approx 8.0$, and $\alpha \approx 30°$.

**Solution:** From the law of sines,

$$\frac{\sin \beta}{8.0} \approx \frac{\sin 30°}{12}$$

From this,

$$\sin \beta \approx 0.33$$

so that

$$\beta \approx 19°$$

While this is the ambiguous case again, we see that the second solution of $\sin \beta = 0.33$, namely $\beta_1 \approx 180° - 19° = 161°$, is inadmissible, as the sum of $\alpha$ and $\beta_1$ would then exceed 180°. It is left to the student to complete the solution of the triangle. ANS. $\gamma \approx 131°$, $c \approx 18$.

## *EXERCISES*

In the following six exercises solve the triangle $ABC$ for the parts requested. Assume that the given measures are approximate and are correct only to the digits given. Indicate by the number of digits retained the accuracy of your results. Use logarithms as directed by the instructor.

**1.** $a = 45.90$, $B = 39°8'$, $A = 57°51'$. Find $C$ and $c$.
**2.** $b = 49.50$, $\beta = 97°23'$, $\gamma = 56°17'$. Find $\alpha$ and $a$.
**3.** $c = 354.0$, $\alpha = 41°36'$, $\beta = 67°0'$. Find $\gamma$ and $b$.
**4.** $\alpha = 90°0'$, $\beta = 27°34'$, $c = 191.6$. Find $a$.
**5.** $\beta = 117°34'$, $\gamma = 34°52'$, $b = 288.1$. Find $\alpha$ and $c$.
**6.** $A = 57°35'$, $B = 64°51'$, $b = 643.5$. Find $C$, $a$, and $c$.

In Exercises **7** through **11**, solve each triangle $ABC$ for all parts not given, if a triangle is determined. If two triangles are possible, find both solutions. Again assume that the measures are approximate, and give results in kind.

**7.** $a = 32$, $b = 49$, $A = 147°$
**8.** $a = 17$, $c = 14$, $C = 82°$
**9.** $a = 21$, $c = 15$, $\alpha = 51°$
**10.** $a = 58$, $b = 37$, $\beta = 64°$
**11.** $c = 8.0$, $a = 10$, $C = 53°$

**12.** Determine the number of triangles possible from the data given for each of the following.

(**a**) $a = 8$, $b = 12$, $A = 45°$
(**b**) $a = 12$, $b = 8$, $\alpha = 60°$
(**c**) $a = 10$, $c = 5$, $C = 120°$
(**d**) $b = 4$, $c = 7$, $\beta = 30°$
(**e**) $c = 8$, $a = 6$, $C = 150°$
(**f**) $b = 5$, $c = 7$, $\beta = 45°$

**13.** From the top of a house 42 ft high the angle of elevation of the top of a tower is 14°10′. From the base of the house the angle of elevation of the top of the tower is 28°20′. Find the height of the tower.

**14.** To find the distance from a point $A$ to an inaccessible point $C$, a base line segment $\overline{AB}$ is laid off. By measurement it is found that $AB \approx 940$ ft, $\angle ABC \approx 42°30'$, $\angle BAC \approx 96°50'$. Find $AC$.

**15.** A pilot flies his plane on a course of 30° from point $A$ to point $B$ 600 miles from $A$. He then flies a course of 150° to a point $C$ that is 600 miles from $A$, his starting point. Find the direction of $A$ from $C$ and the distance $BC$.

**16.** A television tower 120 ft tall stands on the top of a building. From a point on level ground, the angles of elevation of the top and base of the tower are 36°12′ and 27°20′, respectively. Find the height of the building.

---

## 8.3 THE LAW OF COSINES

In triangle $ABC$,

$$a^2 = b^2 + c^2 - 2bc \cos \alpha$$

**PROOF**

Apply a coordinate system to triangle $ABC$ so that $\alpha$ is in standard position with $B$ on the positive $X$ axis, as shown for different possibilities for $\alpha$ in

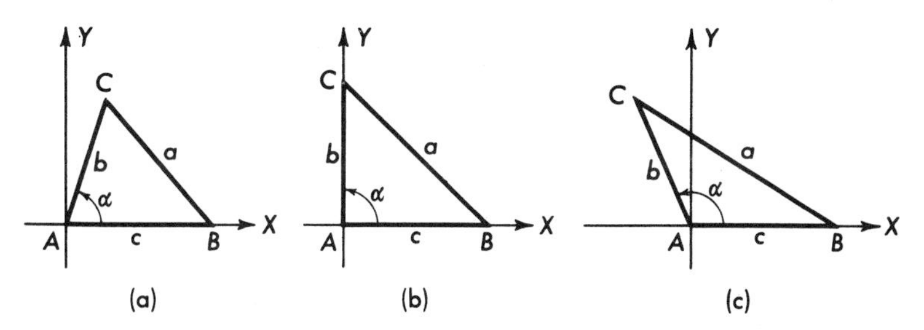


Fig. 8.4 (a) $\alpha$ acute (b) $\alpha$ a right angle (c) $\alpha$ obtuse

Figure 8.4. The vertices then have the coordinates

$$A(0, 0), \quad B(c, 0), \quad C(b \cos \alpha, b \sin \alpha)$$

whether $\alpha$ is acute, obtuse, or a right angle.

From formula (1.3), $d = \sqrt{(x_2 - x_1)^2 + (y_2 - y_1)^2}$, express the square of the distance $a$ between the vertices $B$ and $C$ as

$$\begin{aligned} a^2 &= (b \cos \alpha - c)^2 + (b \sin \alpha - 0)^2 \\ &= b^2 \cos^2 \alpha - 2bc \cos \alpha + c^2 + b^2 \sin^2 \alpha \\ &= b^2 (\cos^2 \alpha + \sin^2 \alpha) + c^2 - 2bc \cos \alpha \end{aligned}$$

Since $\cos^2 \alpha + \sin^2 \alpha = 1$,

$$a^2 = b^2 + c^2 - 2bc \cos \alpha \tag{8.2}$$

This is the law of cosines. In words: *In any triangle, the square of (the length of) any side is equal to the sum of the squares of the other two sides minus twice the product of these sides and the cosine of their included angle.*

(*Note 1:* Since *any* vertex of a triangle can be called $A$, and the other vertices $B$ and $C$, note the generalization of the word form of the law of cosines, and the consequence that

$$b^2 = a^2 + c^2 - 2ac \cos \beta$$

and

$$c^2 = a^2 + b^2 - 2ab \cos \gamma$$

*Note 2:* The law of cosines is a generalized Pythagorean relation with a "built-in" adjustment term, $-2bc \cos \alpha$, which is zero when $\alpha$ is $90°$, which properly decreases the value of $a^2$ when $\alpha$ is acute, and which increases it when $\alpha$ is obtuse. Check these statements.)

When solved for $\cos \alpha$, (8.2) gives a second form of the law of cosines,

$$\cos \alpha = \frac{b^2 + c^2 - a^2}{2bc} \tag{8.3}$$

A third form of the law of cosines, suitable for logarithmic computation, is

$$1 + \cos \alpha = \frac{(a + b + c)(b + c - a)}{2bc} \tag{8.4}$$

To derive this form, add 1 to both members of (8.3) to obtain

$$\begin{aligned} 1 + \cos \alpha &= \frac{b^2 + c^2 - a^2}{2bc} + 1 \\ &= \frac{b^2 + c^2 - a^2}{2bc} + \frac{2bc}{2bc} \\ &= \frac{b^2 + 2bc + c^2 - a^2}{2bc} \\ &= \frac{(b + c)^2 - a^2}{2bc} \\ &= \frac{(b + c + a)\,(b + c - a)}{2bc} \end{aligned}$$

If two sides and the included angle of a triangle are known, (8.2) may be used to find the third side. If the three sides are known, (8.3) or (8.4) may be used to find any of the angles. These applications are illustrated by examples below.

**EXAMPLE 1**

A plane with a cruising speed of 120 mph heads in the direction 52° 30′. A 45-mph wind is blowing in the direction 110°. Find the actual speed and direction of the plane's motion.

**Solution:** A figure is constructed (see Figure 8.5) in which vectors $OA$ and $OB$ denote the air velocity of the plane and the velocity of the wind,

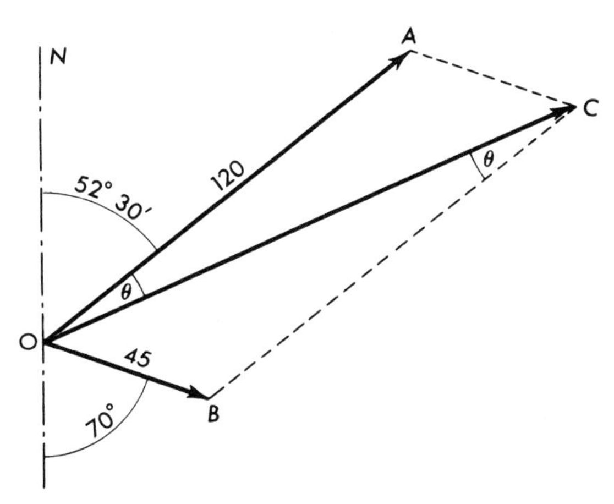


Fig. 8.5 Velocity vectors

respectively. It is required to find the magnitude of the resultant $OC$ and the size of angle $\theta$ which will determine the direction of the resultant. Since the adjacent angles of a parallelogram are supplementary, angle $OBC$ contains 122° 30′. From form (8.2) of the law of cosines as applied to triangle $OBC$,

$$\begin{aligned} OC^2 &= BC^2 + OB^2 - 2BC \cdot OB \cdot \cos \beta \\ &= 120^2 + 45^2 - 2(120)(45) \cos 122° 30' \\ &= 14{,}400 + 2025 - (10{,}800)(-\cos 57° 30') \\ &\approx 14{,}400 + 2025 + (10{,}800)(0.5373) \\ &\approx 14{,}400 + 2025 + 5803 \\ &= 22{,}228 \\ OC &\approx \sqrt{22{,}228} \approx 149 \end{aligned}$$

From the law of sines, in triangle $OBC$,

$$\begin{aligned} \frac{\sin \theta}{45} &\approx \frac{\sin 122° 30'}{149} \\ \log \sin \theta &\approx \log 45 + \log \sin 57° 30' - \log 149 \\ &\approx 9.4060 - 10 \end{aligned}$$

Therefore

$$\theta \approx 14° \, 50'$$

Thus

$$\angle NOC \approx 52° \, 30' + 14° \, 50' = 67° \, 20'$$

The conclusion is that the plane actually flies in a direction 67° 20′ east of north with a speed of 149 mph. (*Note:* It is unlikely that the wind and plane speeds and direction of flight remain constant enough to warrant such accuracy of answer. A more realistic conclusion might be that the plane flies in a direction 67° with a speed of approximately 150 mph.)

### EXAMPLE 2

Two sides of a triangle are 51 and 65. What must their included angle be in order that the third side shall be 20?

**Solution:** Let $\alpha$ be the included angle. Then, by (8.3),

$$\cos \alpha = \frac{51^2 + 65^2 - 20^2}{2(51)(65)} \approx 0.97$$

$$\alpha \approx 14°$$

### EXAMPLE 3

The sides of a triangular field are 4920 ft, 3824 ft, and 5237 ft, respectively. Find the smallest angle of the field.

**Solution:** If $\alpha$ represents the smallest angle, $a = 3824$, because the smallest angle is opposite the shortest side. Thus, by (8.4),

$$1 + \cos \alpha = \frac{(4920 + 3824 + 5237)(4920 + 5237 - 3824)}{2(4920)(5237)}$$

$$= \frac{(13981)(6333)}{(9840)(5237)}$$

Hence

$$\begin{aligned} \log (1 + \cos \alpha) &= \log 13981 + \log 6333 - (\log 9840 + \log 5237) \\ &\approx 4.1455 + 3.8016 - (3.9930 + 3.7191) \\ &= 0.2350 \end{aligned}$$

Therefore

$$1 + \cos \alpha \approx 1.718$$

Thus

$$\cos \alpha \approx 0.718$$

and

$$\alpha = 44° \, 10', \text{ to the nearest 10 min}$$

## EXERCISES

In Exercises **1**, **2**, **3**, **4**, and **5** solve the triangles for the parts requested.

**1.** Given $a = 14.37$, $b = 11.41$, and $C = 52°21'$, find $A$ and $c$.
**2.** Given $b = 729.6$, $c = 613.8$, and $A = 32°12'$, find $B$, $C$, and $a$.
**3.** Given $c = 101.5$, $a = 99.37$, and $B = 47°48'$, find $C$, $A$, and $b$.
**4.** Given $a = 7$, $b = 9$, $c = 12$, find the smallest angle of $ABC$.
**5.** Given $a = 24$, $b = 25$, and $c = 7$, find the largest angle of $ABC$.
**6.** Given the two sides of a triangle as 24 and 16 respectively, and the included angle as 120°. Find the third side. Find the area. (See Exercise 9, Section 5.5.)
**7.** A plane is flying at 200 mph in the direction 60°. A wind is blowing in the direction 300° at 25 mph. Find the plane's ground speed, and direction of flight.
**8.** Two sides of a triangle are 36 and 44, respectively, and the area of the triangle is 396 sq units. Find the angle between the two given sides. Find also the third side of the triangle.
**9.** Two forces of 120 lb and 160 lb respectively act on an object at an angle of 46°20′. Find the magnitude of the resultant.
**10.** Two forces of 40 lb and 60 lb, respectively, acting on an object have a resultant of 75 lb. Find the angle between the two forces.
**11.** Find the distance between the points whose polar coordinates are $(5, \pi/2)$ and $(3, \pi/6)$. [*Hint:* Apply the law of cosines to the triangle determined by these points and the pole, $(0, 0)$.]
**12.** Prove that the distance between the points $P_1$ $(\rho_1, \theta_1)$ and $P_2$ $(\rho_2, \theta_2)$ is given by the formula

$$d = \sqrt{\rho_1{}^2 + \rho_2{}^2 - 2\rho_1\rho_2 \cos (\theta_2 - \theta_1)}$$

**13.** Use the formula of Exercise 12 to find the distance between the points

(**a**) $(3, \pi/2)$ and $(5, \pi/6)$
(**b**) $(2\sqrt{2}, 3\pi/4)$ and $(-\sqrt{2}, 0)$
(**c**) $(10, 56°)$ and $(4, 176°)$
(**d**) $(5, 29°)$ and $(6, -31°)$

**14.** Find the area of the triangle determined by the pole and the pair of points in each of the parts of Exercise 13.
**15.** *A mettle tester.* Sometimes one is "lucky" enough to get the correct answer to a problem after an error has been made. For example, a student reduced the fraction 16/64 to the value ¼ by the unjustified cancellation 1~~6~~/~~6~~4 = ¼. A "favorite" error made in using the law of cosines $a^2 = b^2 + c^2 - 2bc \cos A$ is to multiply $\cos A$ by the entire quantity $b^2 + c^2 - 2bc$, contrary to the formula. Prove that one will never obtain

the correct value of $a$ by accident after making this error. Assume that $a$, $b$, and $c$ *are* sides of a triangle and that no other errors are made.

**16.** *Another mettle tester*. Solve $\triangle ABC$, given $b = 5$, $\beta = 45°$, and $h = 6$, where $h$ is the length of the altitude from $\angle B$.

---

# 8.4 OTHER FORMULAS USED IN TRIANGLE SOLUTION

## A. THE LAW OF TANGENTS

In triangle $ABC$,

$$\frac{\tan \frac{1}{2}(\alpha - \beta)}{\tan \frac{1}{2}(\alpha + \beta)} = \frac{a - b}{a + b} \tag{8.5}$$

This is known as the Law of Tangents. It will be shown that it is useful in the logarithmic determination of the unknown angles of a triangle in which two sides and the included angle are known. An optional form may be derived by using the fact that $\alpha + \beta + \gamma = 180°$, so that

$$\tfrac{1}{2}(\alpha + \beta) = 90° - \tfrac{1}{2}\gamma,$$

and thus

$$\begin{aligned} \tan \tfrac{1}{2}(\alpha + \beta) &= \tan (90° - \tfrac{1}{2}\gamma) \\ &= \cot \tfrac{1}{2}\gamma \end{aligned}$$

Hence we may write (8.5) as

$$\tan \tfrac{1}{2}(\alpha - \beta) = \left(\frac{a - b}{a + b}\right) \cot \tfrac{1}{2}\gamma$$

### PROOF OF THE LAW OF TANGENTS

From the law of sines,

$$\frac{\sin \alpha}{\sin \beta} = \frac{a}{b}$$

Hence*

$$\frac{\sin \alpha - \sin \beta}{\sin \alpha + \sin \beta} = \frac{a - b}{a + b}$$

---

* If $\frac{a}{b} = \frac{c}{d}$, then $\frac{a - b}{a + b} = \frac{c - d}{c + d}$. PROOF: Subtract 1 from both sides of the first equation and obtain $\frac{a - b}{b} = \frac{c - d}{d}$. Now add 1 to both sides of the same equation and obtain $\frac{a + b}{b} = \frac{c + d}{d}$. Dividing equals by equals gives $\frac{a - b}{a + b} = \frac{c - d}{c + d}$.

Application of identities (6.27) and (6.26) to the numerator and denominator of the left member gives

$$\frac{2 \cos \frac{1}{2}(\alpha + \beta) \sin \frac{1}{2}(\alpha - \beta)}{2 \sin \frac{1}{2}(\alpha + \beta) \cos \frac{1}{2}(\alpha - \beta)} = \frac{a - b}{a + b}$$

This readily reduces to the (8.5) form of the law of tangents.

**EXAMPLE**

Two sides of a triangle are 236.7 yd and 341.3 yd, respectively. They form with each other an angle of 67° 40′. Find the other angles of the triangle.

**Solution:** *To avoid negative signs*, let the longer of the two sides correspond to the side $a$ of the law of tangents. If $\gamma = 67° 40'$, $\alpha + \beta = 180° - 67° 40' = 112° 20'$ or $\frac{1}{2}(\alpha + \beta) = 56° 10'$. Now $a - b = 104.6$, and $a + b = 578.0$. Substituting these values in (8.5),

$$\frac{\tan \frac{1}{2}(\alpha - \beta)}{\tan 56° 10'} = \frac{104.6}{578.0}$$

Hence

$$\begin{aligned} \log \tan \tfrac{1}{2}(\alpha - \beta) &= \log \tan 56° 10' + \log 104.6 - \log 578.0 \\ &\approx 0.1737 + 2.0195 - 2.7619 \\ &= 9.4313 - 10 \end{aligned}$$

Therefore

$$\tfrac{1}{2}(\alpha - \beta) \approx 15° 6'$$

But

$$\tfrac{1}{2}(\alpha + \beta) = 56° 10'$$

If these two equalities are added,

$$\alpha \approx 71° 16'$$

If the first is subtracted from the second,

$$\beta \approx 41° 4'$$

If desired, the solution of the triangle for the third side may now be carried out by using the law of sines.

[*Note:* While it is true that the sum of $\alpha$, $\beta$, and $\gamma$ should equal 180°, this does not give a complete check on the accuracy of the work. If $\frac{1}{2}(\alpha + \beta)$ is found correctly and this value is combined correctly with the value found for $\frac{1}{2}(\alpha - \beta)$, the resulting three angles will total 180°, regardless of the correctness of the value found for $\frac{1}{2}(\alpha - \beta)$. Hence this check should be used only to discover errors in finding $\frac{1}{2}(\alpha + \beta)$ and in adding and subtracting the last pair of equations.]

## B. RELATION OF THE HALF-ANGLE TO THE THREE SIDES

The problem of finding an angle of a triangle for which the three sides are known has been solved by the law of cosines. A further modification of the (8.4) form of the law of cosines can be made which yields an even more convenient formula, if a logarithmic solution is desired. Dividing both members of (8.4) by 2,

$$\frac{1 + \cos \alpha}{2} = \frac{(a + b + c)(b + c - a)}{4bc}$$

But, by (6.24),

$$\frac{1 + \cos \alpha}{2} = \cos^2 \frac{\alpha}{2}$$

Therefore

$$\cos \frac{\alpha}{2} = \sqrt{\frac{(a + b + c)(b + c - a)}{4bc}}$$

An abbreviated form of this is obtained by noting that, if $\dfrac{a + b + c}{2} = s$ (the semiperimeter of the triangle), then $\dfrac{b + c - a}{2} = s - a$ (why?), so that

$$\cos \frac{\alpha}{2} = \sqrt{\frac{s(s - a)}{bc}}, \quad s = \frac{a + b + c}{2} \tag{8.6}$$

Observe that there is no choice of sign for this square root because $\alpha < 180°$; therefore $\frac{\alpha}{2} < 90°$ and $\cos \frac{\alpha}{2} > 0$.

### EXAMPLE

To permit comparison, solve Example 3, Section 8.3, with the aid of (8.6). The problem is to find the smallest angle of a triangle whose sides are 4920, 3824, and 5237, respectively.

**Solution:** It is known that

$$s = \frac{4920 + 3824 + 5237}{2} = 6990.5$$

Thus, if $a = 3824$, the shortest side, then $s - a = 3166.5$. Hence

$$\cos \frac{\alpha}{2} = \sqrt{\frac{s(s - a)}{bc}}$$

$$= \sqrt{\frac{(6990.5)(3166.5)}{(4920)(5237)}}$$

Now

$$\log \cos \frac{\alpha}{2} = \tfrac{1}{2}[\log 6990.5 + \log 3166.5 - (\log 4920 + \log 5237)]$$
$$\approx \tfrac{1}{2}[3.8455 + 3.5006 - (3.6920 + 3.7191)]$$
$$= \tfrac{1}{2}(9.9340 - 10)$$
$$= 4.9670 - 5$$

Therefore

$$\frac{\alpha}{2} = 22^\circ\ 4', \text{ approximately correct to the nearest minute.}$$

Hence

$$\alpha = 44^\circ\ 8', \text{ to the nearest 2 min.}$$

## *EXERCISES*

Assume all given measures are approximate.

In Exercises **1** through **5** use the law of tangents to find the unknown angles in the triangle $ABC$.

**1.** $a = 247.5$, $b = 194.5$, $C = 38°30'$
**2.** $a = 47.3$, $b = 83.4$, $C = 49°40'$
**3.** $a = 86$, $c = 47$, $B = 18°$
**4.** $b = 4784$, $c = 5341$, $A = 71°36'$
**5.** $c = 37.40$, $a = 56.82$, $B = 53°28'$

In Exercises **6** through **10** use the cosine of the half-angle formula (8.6) to find the angle indicated in the problem.

**6.** $a = 47$, $b = 85$, $c = 62$, find $A$.
**7.** $a = 375$, $b = 292$, $c = 483$, find $B$.
**8.** $a = 284.1$, $b = 354.4$, $c = 402.3$, find $C$.
**9.** $a = 48.47$, $b = 38.93$, $c = 52.12$, find the largest angle.
**10.** In Exercise **8** find angles $A$ and $B$ independently and check that the sum of the measures of the three angles is approximately 180°.
**11.** A pilot flies his plane on a course 14°10′ from $A$ to $B$ in two hours at 200 mph. He then flies a course 297°15′ for 3 hr at the same speed. Find the direction of $C$ from $A$, and also the distance $AC$.
**12.** A plane flies from $P$ to $Q$ in 3 hr. The plane heads northeast at 200 mph air speed. If the direction of $Q$ from $P$ is 49°, find the wind direction and wind speed if $PQ$ is 650 miles.
**13.** The diagonals of a parallogram intersect at an angle of 80°. If the diagonals are 60 ft and 70 ft, respectively, find the angles formed by the sides of the parallegram.

**14.** Two forces acting from a common point have magnitudes of 100 lb and 80 lb, respectively. A third force of 60 lb acting in the same plane keeps these forces in equilibrium. Find the angle between the 100 lb and 80 lb forces. *Hint:* The third force must be equal in magnitude but opposite in direction to the resultant of the first two forces.

**15.** A pilot wishes to fly due east. In what direction must he head his plane if a south wind is blowing at 20 mph and his air speed is 300 mph. Find also his ground speed.

## 8.5 CYCLICAL PERMUTATIONS IN THE TRIANGLE FORMULAS

Any of the triangle formulas may be changed to an equally valid formula by replacing each letter of the formula by the letter next in succession to it in such a circular arrangement as shown in Figure 8.6. Such a circular rear-

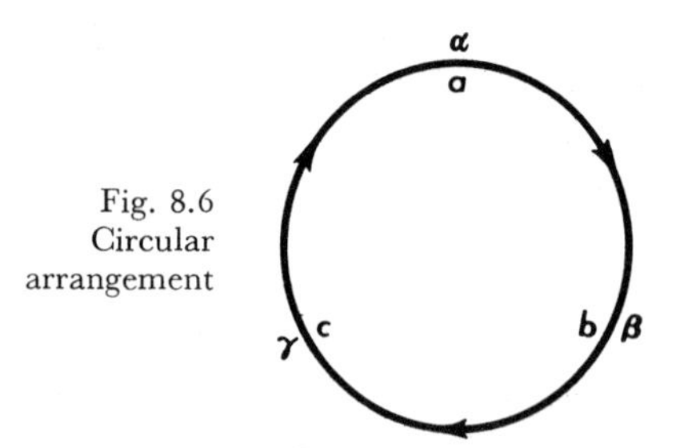


Fig. 8.6 Circular arrangement

rangement of letters is called a cyclical permutation of the letters. As an example, consider the formula for the area $K$ of triangle $ABC$ developed as an exercise in Chapter 4,

$$K = \tfrac{1}{2}bc \sin \alpha$$

A cyclical permutation yields the formula

$$K = \tfrac{1}{2}ca \sin \beta$$

A second cyclical change gives

$$K = \tfrac{1}{2}ab \sin \gamma$$

Another such change restores the original formula.

The idea of cyclical permutation in the triangle formulas has been introduced largely as a matter of mathematical interest. It should supplement, not displace, the mastering of the relationship given by the formula. The laws do not depend upon the notation used in describing the triangle.

### SUMMARY OF FORMULAS FOR SOLUTION OF TRIANGLES

*The law of sines:*

$$\frac{a}{\sin \alpha} = \frac{b}{\sin \beta} = \frac{c}{\sin \gamma} \tag{8.1}$$

*The law of cosines:*

$$a^2 = b^2 + c^2 - 2bc \cos \alpha \tag{8.2}$$

$$\cos \alpha = \frac{b^2 + c^2 - a^2}{2bc} \tag{8.3}$$

$$1 + \cos \alpha = \frac{(a + b + c)(b + c - a)}{2bc} \tag{8.4}$$

*The law of tangents:*

$$\frac{\tan \frac{1}{2}(\alpha - \beta)}{\tan \frac{1}{2}(\alpha + \beta)} = \frac{a - b}{a + b} \tag{8.5}$$

*Cosine of the half-angle formula:*

$$\cos \frac{\alpha}{2} = \sqrt{\frac{s(s - a)}{bc}} \tag{8.6}$$

where

$$s = \frac{a + b + c}{2}$$

## MISCELLANEOUS EXERCISES/CHAPTER 8

1. A boat is running in a northerly direction at a speed of 20 knots. A lighthouse has a bearing of 21° 25′ at 7 PM, and at 9:30 PM the bearing of the lighthouse is 42°20′. At what time and at what distance will the boat pass closest to the lighthouse?
2. A tower 80 ft in height stands on a hill. The angles of elevation of the bottom and top of the tower from a point on an island are 8°15′ and 10°35′ respectively. Find the distance from the bottom of the tower to the point on the island.
3. A plane is heading in a direction 310° at 250 mph. If there is a 30 mph wind blowing in the direction 30°, find his ground speed and course of flight.

4. A farmer has a triangular field. If two sides are 1000 yd and 2400 yd respectively, and the angle between the two given sides is 100°, find the area.
5. From a point in front of a building the angles of elevation of the top and bottom of a 40 ft flag pole on top of the building are 35°17′ and 28°57′ respectively. Find the height of the building above the point.
6. A 30.06 ft pole is standing vertically on a hill whose slope is 15° with the horizon and sloping away from the sun at a given time. Find the angle of elevation of the sun if the pole's shadow on the slope is 44.3 ft in length.
7. Two observers at $A$ and $B$ observe a flash of lightning at $C$. The noise of the thunder gets to $A$ in 3 sec, and to $B$ in 4 sec. If $AB =$ 2000 yd find angles $CAB$ and $CBA$, assuming sound travels at 1100 fps and neglecting the speed of light. (*Hint:* use cosine of the half-angle formula.)
8. Three circles whose radii are 3, 4, and 5 are tangent externally. Find the area enclosed between them.
9. *A mettle tester*. Two sides of a triangle are 60 and 80, and the difference of their opposite angles is 30°. Find the third side.
10. A flag pole 50 ft long is painted red and white. The red part is 20 ft. How far from the foot of the pole will the two parts subtend equal angles? Assume the red part to be on the bottom of the pole.
11. *A mettle tester*. Prove Heron's rule that the area $K$ of triangle $ABC$ is given by the formula

$$K = \sqrt{s(s - a)(s - b)(s - c)}$$

where $s = \dfrac{a + b + c}{2}$. [*Hint:* In the same way that formula (8.4) for $1 + \cos \alpha$ was obtained, derive a formula for $1 - \cos \alpha$. From the products of the corresponding members of these formulas, obtain a formula for $\sin^2 \alpha$. Finally, apply this to the area formula

$$K = \tfrac{1}{2}bc \sin \alpha$$

and simplify the result.]
12. In surveying a tract of land an engineer found it impracticable to measure the side $AB$ (see the figure) because of a thick brushwood lying between $A$ and $B$. He therefore measured $AE$, 9.17 chains, and $EB$, 3.12 chains, and found the angle at $E = 73°\ 7'$. Find the distance from $A$ to $B$.

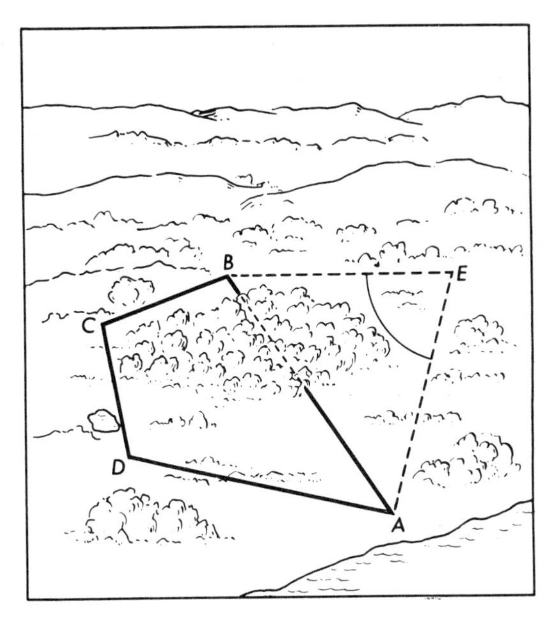


Ex. 12

**13.** A pilot must fly his plane from $A$ 500 mi to $B$ and return. The bearing of $B$ from $A$ is 120°. A 20 mph wind is blowing in the direction 150°. If his air speed is 200 mph, find his heading of flight from $A$ to $B$, and his heading of flight from $B$ to $A$, and also the total time for the round trip.

**14.** A triangle has sides of 343, 297, and 410. Find the largest angle and the area of the triangle.

**15.** A boat whose speed in still water is 4 mph moves straight across a river 3,000 ft wide. If the current is 3 mph, find the heading the boat must take, and also the time in crossing the river.

**16.** What heading should the boat take in Exercise **15** if the point of landing is 100 ft downstream?

**17.** A force of 60 lb has a direction of 120°, and another force of 100 lb acting from the same point has a direction of 200°. Find the magnitude and the direction of the resultant.

**18.** Show that the diameter of the circumscribed circle of triangle $ABC$ is equal to $b \csc B$.

**19.** A person is on a bank of a lake, the slope of the bank with the horizontal being 20°25′. At a point 100 ft from the water's edge the angle of depression of the water's edge across the lake is 2°15′. Find the width of the lake.

**20.** Find the angle of elevation of the sun when the longest shadow that a straight rod can cast is five times the length of the rod itself. (*Note:* The longest shadow is not cast by the rod when in a vertical position.)

**21.** A plot of ground is in the shape of a quadrilateral. The lengths of the sides are: $a = 300$ yd, $b = 475$ yd, $c = 525$ yd, and $d = 600$ yd. If the angle between $a$ and $b$ is 50°, find the area of the plot.

**22.** *A mettle tester*. A person travelling 900 yd towards a distant tower notes that the angle of elevation of the top of the tower is doubled. Going 600 yd more toward the tower the angle of elevation has again doubled. Find the height of the tower.

# CHAPTER 9.
# COMPLEX NUMBERS

## 9.1 COMPLEX NUMBERS

The formal solution of the equation

$$x^2 = -9$$

leads to the result

$$x = \pm \sqrt{-9}$$

The fact that there is no real number whose square equals $-9$ causes us to call expressions like $\sqrt{-9}$ *imaginary numbers*. The term *imaginary* must not be taken to mean *useless*, for the practical applications of these numbers are widespread and important.

We assume that

$$\begin{aligned}\sqrt{-9} &= \sqrt{-1} \cdot \sqrt{9} \\ &= 3\sqrt{-1}\end{aligned}$$

The imaginary number $\sqrt{-1}$ is called the *imaginary unit* and is denoted by the symbol $i$. Note that, if

$$i = \sqrt{-1}$$

then, by the definition of a square root,

$$i^2 = -1$$

In general, for every positive number $p$,

$$\begin{aligned}\sqrt{-p} &= \sqrt{-1} \cdot \sqrt{p} \\ &= i\sqrt{p}\end{aligned}$$

A *complex number* is a number of the form $a + bi$, where $a$ and $b$ are real. The number $a$ is called the *real part*, and $b$ the *imaginary part*. Every real number may be expressed as a complex number in which $b = 0$. If $a = 0$,

$b \neq 0$, the number is called a *pure imaginary* number. As examples, the number 5 may be expressed as $5 + 0i$, and $\sqrt{-5}$ as $0 + i \cdot \sqrt{5}$. The set of real numbers is a subset of the set of complex numbers.

Two complex numbers $a + bi$ and $c + di$ are equal if and only if $a = c$ and $b = d$. If two complex numbers are of the form $a + bi$ and $a - bi$, each is called the *conjugate* of the other. For example, the conjugate of $2 - 3i$ is $2 + 3i$.

A logical foundation for the mathematical operations with complex numbers (as well as with the real numbers) is beyond the scope of this book. We content ourselves with pointing out that, if $i$ is subject to the same rules that govern the fundamental operations of ordinary algebra, the results will be consistent. Thus, starting with

$$i^2 = -1$$

we have

$$i^3 = i^2 \cdot i = -i$$
$$i^4 = (i^2)^2 = (-1)^2 = 1$$

The odd powers of $i$ will equal either $i$ or $-i$. The even powers will equal either $-1$ or 1. For example,

$$\begin{aligned} i^{54} &= i^{4\times13+2} \\ &= (i^4)^{13} \cdot (i)^2 \\ &= 1^{13} \cdot (-1) \\ &= -1 \end{aligned}$$

Other operations with complex numbers are illustrated by examples below. Note how $i$ is treated *as though* it were a real number with the added property that $i^2 = -1$.

**EXAMPLE 1**

$$\begin{aligned} \sqrt{-12} \cdot \sqrt{-3} &= i\sqrt{12} \cdot i\sqrt{3} \\ &= i^2\sqrt{36} \\ &= -6 \end{aligned}$$

(*Note:* Pure imaginaries do not follow the rule for real numbers that

$$\sqrt{a} \cdot \sqrt{b} = \sqrt{ab}$$

employed in getting

$$\sqrt{12} \cdot \sqrt{3} = \sqrt{36}$$

for its use with pure imaginaries would give such results as $\sqrt{-12} \cdot \sqrt{-3}$ falsely being set equal to $\sqrt{(-12)(-3)} = \sqrt{36} = 6$, as compared with the correct value $-6$ found above. This is an important reason why the notation

$i\sqrt{3}$ is preferred to $\sqrt{-3}$, because the real coefficient of $i$, namely $\sqrt{3}$, behaves as real numbers are known to behave. The use of this notation, together with the rule of treating $i$ as a real number with the property that $i^2 = -1$, will assure that pitfalls of the above-mentioned type are avoided.)

**EXAMPLE 2**

$$(5 - 3i) - (7 + 4i) = -2 - 7i$$

**EXAMPLE 3**

$$\begin{aligned}(5 - \sqrt{-9})(7 + \sqrt{-16}) &= (5 - 3i)(7 + 4i)\\ &= 35 - i - 12i^2\\ &= 35 - i + 12\\ &= 47 - i\end{aligned}$$

**EXAMPLE 4**

$$\begin{aligned}\frac{5 - 3i}{7 + 4i} &= \frac{(5 - 3i)(7 - 4i)}{(7 + 4i)(7 - 4i)}\\ &= \frac{23 - 41i}{65}\\ &= \frac{23}{65} - \frac{41}{65}i\end{aligned}$$

(Note that the numerator and the denominator of the given fraction are multiplied by the conjugate of the denominator.)

**EXAMPLE 5**

Solve for $x$ and $y$, assuming that they are real:

$$x + (2 - y)i = (3 + y) - 2xi$$

**Solution:** Applying the definition of the equality of two complex numbers, equate the real parts of the two members to obtain the equation

$$x = 3 + y$$

and the imaginary parts to obtain

$$2 - y = -2x$$

The simultaneous solution of this pair of equations yields

$$x = -5, \quad y = -8$$

## EXERCISES

In Exercises **1** to **14** perform the indicated operations and then put each result in the form $a + ib$.

**1.** $(3 + 2i) + (5 - 2i)$

**2.** $(2i - 3) + (3 + 2i)$

**3.** $(4 + i) + (3 - i) + (i - 7)$

**4.** $(3 - 2i) + (5 + 4i)$

**5.** $(6 + 2i) - (2 + 3i)$

**6.** $4 - (4 - 2i)$

**7.** $2i - (3 + i)$

**8.** $(2 + 3i) - (4 - 4i) + (i - 1)$

**9.** $(2 - 3i)^2$

**10.** $(2 + i)(2 - i)$

**11.** $(3 - i)(4 + 2i)(3 + i)$

**12.** $\dfrac{(3 + 2i)}{(4 - 2i)}$

**13.** $\dfrac{2}{i}$

**14.** $\dfrac{(2 + 3i)}{(1 + 2i)}$

**15.** Prove that the sum and product of two conjugate complex numbers are real.

In Exercises **16** to **20** solve for $x$ and $y$, assuming that they are real.

**16.** $2x + 3i = 2 + iy$

**17.** $x + 2y + 2iy = 3 + 4i$

**18.** $2x - 3y + (x + 2y)i = 4 - 5i$

**19.** $x - xyi = y - 4i$

**20.** $x^2 + (y - 1)i = 25 - y^2 + xi$

**21.** Find the square roots of $i$. [*Hint:* Assume that a square root of $i$ is the number $x + yi$. Then

$$(x + yi)^2 = i]$$

**22.** If $\alpha = -\frac{1}{2} + \dfrac{i\sqrt{3}}{2}$ and $\beta = -\frac{1}{2} - \dfrac{i\sqrt{3}}{2}$, show that

$$\alpha^2 = \beta \quad \text{and} \quad \beta^2 = \alpha$$

**23.** If $\alpha$ and $\beta$ are as given in Exercise **22**, show that $\alpha^3 = \beta^3 = 1$.

## 9.2 GRAPHICAL REPRESENTATION OF COMPLEX NUMBERS. POLAR FORM

The complex number $x + yi$ may be represented graphically by the point $(x, y)$ plotted on a rectangular coordinate system. To each such complex number there corresponds a unique point of the plane, and, conversely, to each point of the plane there corresponds a unique complex number. All the points on the $X$ axis represent the real numbers $x + 0i$, while the points on the $Y$ axis represent the pure imaginaries $0 + yi$. Hence the $X$ axis is called the *axis of reals*, and the $Y$ axis, the *axis of imaginaries*. The plane is referred to as the *complex plane*.

If $Z$ is the complex point $x + iy$, the length $r$ of the radius vector $AZ$ is called the *modulus or absolute value* of the number $x + iy$. The angle $\theta = \angle XOZ$ is called the *amplitude or argument* of $x + yi$. (See Figure 9.1.) These are referred to as mod $Z$ or $|Z|$, and as amp $Z$ or arg $Z$, respectively.

We note that the amplitude of a given complex number is not unique. If $\theta$ is the amplitude of a given complex number, then any angle coterminal with $\theta$ is also an amplitude of the same number. That is, $\theta + n \cdot 2\pi$, for $n =$ any integer, is an amplitude if $\theta$ is measured in radians, or $\theta + n \cdot 360°$, if $\theta$ is in degrees. This reminds us of the multiplicity of choices of $\theta$ for a point

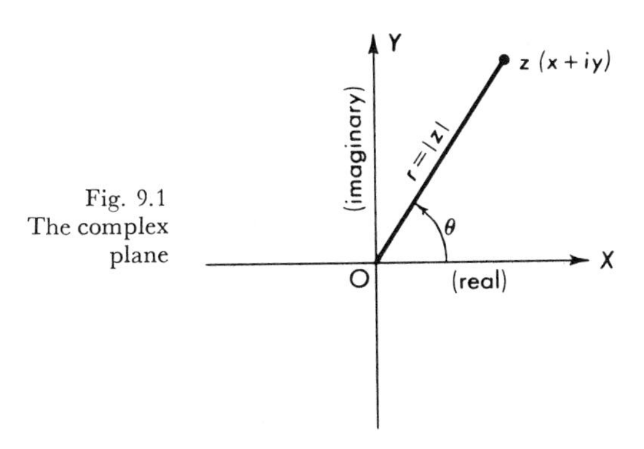


Fig. 9.1
The complex plane

in polar coordinates. We arbitrarily select the smallest non-negative value of $\theta$ as the principle amplitude of a given complex number.

In polar form, the number $x + iy$ corresponds to the point $(r, \theta)$. Note that, by definition, $r = |OZ|$, so that $r$ differs from the $\rho$ of ordinary polar coordinates in that $\rho$ may be negative, while $r$ is positive or zero. Using the fact that

$$x = r \cos \theta$$

and

$$y = r \sin \theta$$

write

$$x + yi = r \cos \theta + ir \sin \theta$$

so that

$$x + yi = r (\cos \theta + i \sin \theta)$$

The expression $r (\cos \theta + i \sin \theta)$ is called the *polar form* of $x + iy$.*

### EXAMPLE 1

Plot the point $3 + 3i$. Find the modulus and smallest positive amplitude of the number $3 + 3i$. Express the number in polar form.

---

* The polar form $r (\cos \theta + i \sin \theta)$ is conveniently expressed by the notation $r$ cis $\theta$; c for cosine, i for $\sqrt{-1}$, and s for sine. We shall employ this notation presently.

**Solution:** The point $3 + 3i$, with coordinates $(3, 3)$ is shown in Figure 9.2.

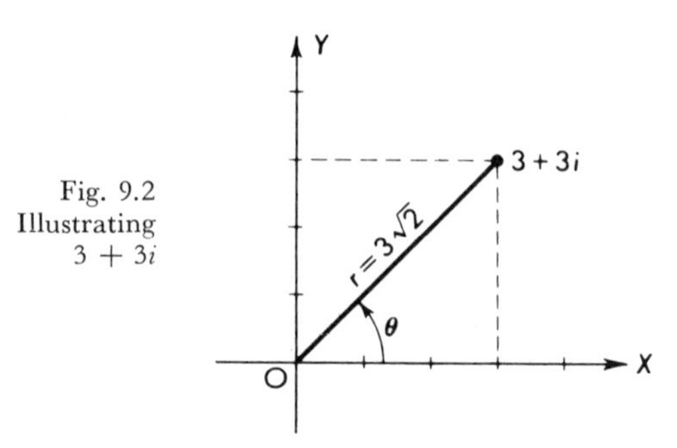


Fig. 9.2 Illustrating $3 + 3i$

By inspection (or by noting that $\tan\theta = 3/3 = 1$), it is found that $\theta = \text{amp}\,(3 + 3i) = 45°$. Also,

$$r = \text{mod}\,(3 + 3i) = \sqrt{3^2 + 3^2} = 3\sqrt{2}.$$

The polar form is

$$r\,(\cos\theta + i\sin\theta) = 3\sqrt{2}\,(\cos 45° + i\sin 45°).$$

### EXAMPLE 2

Express the number $2 - 3i$ in polar form.

**Solution:** The point $(2, -3)$ lies in the fourth quadrant, so that $\theta = \arctan(-3/2)$ must be chosen to fall in the fourth quadrant. From the table of natural functions (Table III), $\tan 56°\,19' = 3/2$, so that the reference angle is $56°\,19'$. Hence $\theta = -56°\,19'$, or, if a positive amplitude is desired, $\theta = 360° - 56°\,19' = 303°\,41'$. Also, $r = \sqrt{2^2 + 3^2} = \sqrt{13}$. Hence

$$\begin{aligned} 2 - 3i &= r\,(\cos\theta + i\sin\theta) \\ &= \sqrt{13}\,(\cos 303°\,41' + i\sin 303°\,41') \end{aligned}$$

A general polar form is

$$2 - 3i = \sqrt{13}\,[\cos\,(303°\,41' + n\cdot 360°) + i\sin\,(303°\,41' + n\cdot 360°)]$$

where $n$ is any integer. Of course, $303°\,41'$ is approximate.

### EXAMPLE 3

Express $-4$ in general polar form, using radian angle measure.

**Solution:** By inspection, the point $(-4, 0)$ is located on the $\theta = \pi$ line and at a distance of 4 units from the origin. Hence $\text{amp}\,(-4) = \pi + n\cdot 2\pi$, and $\text{mod}\,(-4) = 4$, so that

$$-4 = 4\,[\cos\,(\pi + n\cdot 2\pi) + i\sin\,(\pi + n\cdot 2\pi)]$$

or

$$-4 = 4\,[\cos\,(2n + 1)\pi + i\,\sin\,(2n + 1)\pi]$$

Given two complex numbers, $A = a + bi$ and $B = c + di$; if $C = A + B$,

$$C = (a + c) + (b + d)i$$

Graphically, the points $A$, $B$, and $C$ are so related that the figure $AOBC$ is a parallelogram (see Figure 9.3). (It is left to the student to prove this by

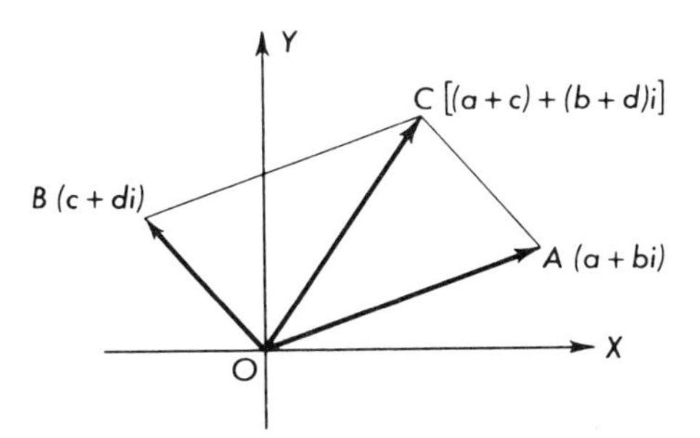


Fig. 9.3 Addition of complex numbers

using the distance formula to show that $OA = BC$ and that $OB = AC$.) The fact that the addition of complex numbers thus follows the parallelogram law for the addition of vectors suggests that complex numbers may be represented by vectors: the complex number $A = a + bi$, by the vector **OA,** and so on. Conversely, any vector **OP** can be represented by the complex number $x + yi$, where $(x, y)$ is the point $P$.

The graphical subtraction of $B = c + di$ from $A = a + bi$ is performed by noting that

$$\begin{aligned} A - B &= (a + bi) - (c + di) \\ &= (a + bi) + (-c - di) \end{aligned}$$

and proceeding with the graphical addition of $A$ and $-B$ as shown in Figure 9.4.

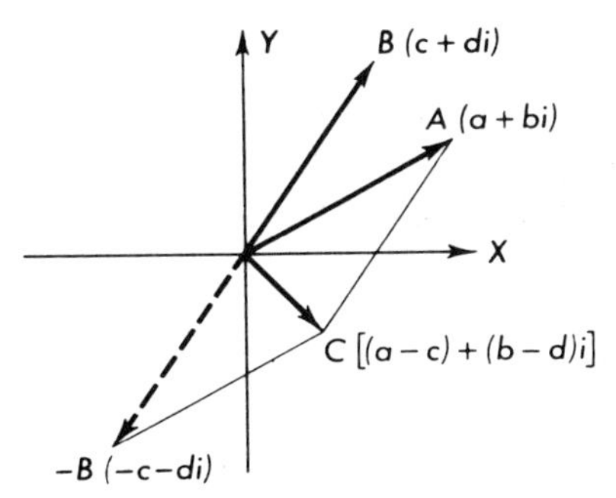


Fig. 9.4 Subtraction of complex numbers

## EXERCISES

In Exercises **1** to **8** plot the complex number, find its modulus, its smallest non-negative amplitude, and write the number in polar form.

**1.** $4 + 3i$
**2.** $3 - 4i$
**3.** $-5 + 12i$
**4.** $3i$
**5.** $-4$
**6.** $3i - 6$
**7.** $\sqrt{3} - 2i$
**8.** $\cos \pi/3 - i \sin \pi/3$
**9.** Compare the definition of the ordinary absolute value of a real number with that of the absolute value of the complex number, and show that the former is a special case of the latter.

In Exercises **10** to **19** perform the indicated operation graphically and then check each one algebraically.

**10.** $(3 + 4i) + (2 - 3i)$
**11.** $(3 - 4i) - (2 + 3i)$
**12.** $3 + (4 - 2i)$
**13.** $(2i - 1) - (4i + 3)$
**14.** $(2 + 3i) + (6 - 4i) + (-2 + i)$
**15.** $2i - (3 - 5i)$
**16.** $(3 + 0i) - (0 + 4i)$
**17.** $(2 - 3i) + (-1 + i)$
**18.** $(2 + i) + (3 - i)$
**19.** $(4 + 2i) - (4 - 2i)$

In Exercises **20** to **25** express each complex number in the form $a + bi$ without trigonometric forms.

**20.** $3(\cos \pi/3 + i \sin \pi/3)$
**21.** $\sqrt{2}(\cos 225° - i \sin 225°)$
**22.** $2(\cos \pi + i \sin \pi)$
**23.** $4(\cos 120° + i \sin 120°)$
**24.** $5[\cos (-60°) + i \sin (-60°)]$
**25.** $6(\cos 5\pi + i \sin 5\pi)$

In the following Exercises find the smallest non-negative value of $\theta$ for which:

**26.** $2(\cos \theta + i \sin \theta) = 1 + i\sqrt{3}$
**27.** $\sqrt{2}\,(\cos \theta + i \sin \theta) = 1 - i$
**28.** $3(\cos \theta + i \sin \theta) = 3[\cos (-225°) + i \sin (-225°)]$

## 9.3 THE PRODUCT OF TWO OR MORE COMPLEX NUMBERS IN THE POLAR FORM. DE MOIVRE'S FORMULA

If two complex numbers,

$$z_1 = x_1 + iy_1 \quad \text{and} \quad z_2 = x_2 + iy_2$$

are expressed in *polar form*, their product is

$$z_1z_2 = r_1 (\cos \theta_1 + i \sin \theta_1) \cdot r_2 (\cos \theta_2 + i \sin \theta_2)$$

By actual multiplication and use of the fact that $i^2 = -1$, this product can be written in the form

$$\begin{aligned} z_1z_2 &= r_1r_2 \,[\cos\theta_1 \cos\theta_2 - \sin\theta_1 \sin\theta_2 + i\,(\sin\theta_1 \cos\theta_2 + \cos\theta_1 \sin\theta_2)] \\ &= r_1r_2\,[\cos\,(\theta_1 + \theta_2) + i \sin\,(\theta_1 + \theta_2)] \end{aligned}$$

Using this result, the product of three complex numbers is easily obtained:

$$z_1z_2z_3 = r_1r_2r_3\,[\cos\,(\theta_1 + \theta_2 + \theta_3) + i \sin\,(\theta_1 + \theta_2 + \theta_3)]$$

When the process is continued, the general result is

$$z_1z_2 \cdots z_n = r_1r_2 \cdots r_n \\ [\cos\,(\theta_1 + \theta_2 + \cdots + \theta_n) + i \sin\,(\theta_1 + \theta_2 + \cdots + \theta_n)]$$

If now the factors in this formula are equal and the form $r\,(\cos\theta + i \sin\theta)$ is used, there results

$$z^n = [r\,(\cos\theta + i\sin\theta)]^n = r^n\,(\cos n\theta + i \sin n\theta)$$

The result,

$$[r\,(\cos\theta + i\sin\theta)]^n = r^n\,(\cos n\theta + i \sin n\theta) \qquad (9.1)$$

is known as De Moivre's formula, after the French-born (but English-educated) mathematician, A. De Moivre. [*Note:* We have proved De Moivre's formula for $n$ a positive integer. It is also true if $n$ is a negative integer (see Exercise 8, Section 9.4) and if $n$ is any rational fraction (Exercise 21, Miscellaneous Exercises, this chapter).]

### EXAMPLE 1

Express $(\sqrt{3} + i)^{10}$ as a complex number.

**Solution:** First express $\sqrt{3} + i$ in the form

$$r(\cos\theta + i \sin\theta)$$

by noting that

$$r = \sqrt{x^2 + y^2} = \sqrt{3 + 1} = 2$$

and that

$$\theta = \mathrm{Tan}^{-1}\left(\frac{y}{x}\right) = \mathrm{Tan}^{-1}\left(\frac{1}{\sqrt{3}}\right)$$

Hence

$$\theta = \frac{\pi}{6}$$

Therefore

$$\sqrt{3} + i = 2\left(\cos\frac{\pi}{6} + i \sin\frac{\pi}{6}\right)$$

By De Moivre's formula,

$$\begin{aligned}(\sqrt{3}+i)^{10} &= 2^{10}\left(\cos\frac{10\pi}{6}+i\sin\frac{10\pi}{6}\right)\\ &= 1024\left(\frac{1}{2}-\frac{i\sqrt{3}}{2}\right)\\ &= 512-512\sqrt{3}\,i\end{aligned}$$

**EXAMPLE 2**

With the aid of De Moivre's formula, express $\cos 3\theta$ and $\sin 3\theta$ in terms of $\sin\theta$ and $\cos\theta$.

**Solution:** By De Moivre's formula,

$$(\cos\theta+i\sin\theta)^3=\cos 3\theta+i\sin 3\theta$$

But, by direct multiplication,

$$(\cos\theta+i\sin\theta)^3=\cos^3\theta+3i\cos^2\theta\sin\theta+3i^2\cos\theta\sin^2\theta+i^3\sin^3\theta$$

Simplifying powers of $i$ and collecting terms of the right member give

$$(\cos\theta+i\sin\theta)^3=\cos^3\theta-3\cos\theta\sin^2\theta+i(3\cos^2\theta\sin\theta-\sin^3\theta)$$

Hence

$$\cos 3\theta+i\sin 3\theta=\cos^3\theta-3\cos\theta\sin^2\theta+i(3\cos^2\theta\sin\theta-\sin^3\theta)$$

Equating real parts and imaginary parts,

$$\cos 3\theta=\cos^3\theta-3\cos\theta\sin^2\theta$$

and

$$\sin 3\theta=3\cos^2\theta\sin\theta-\sin^3\theta$$

An equivalent result was previously obtained by considerably more labor in Exercises 11 and 13 of Part 2 of the Miscellaneous Exercises for Chapter 6.

## *EXERCISES*

**1.** Verify the following statement: The modulus of the product of two complex numbers is the product of their moduli, and the amplitude of their product is the sum of their amplitudes.

In Exercises **2** through **5**, perform the indicated operation by using the product rule, simplifying the result.

**2.** $[2(\cos\pi/3+i\sin\pi/3)][3(\cos\pi/2+i\sin\pi/2)]$
**3.** $[3(\cos 100^\circ+i\sin 100^\circ)][4(\cos 20^\circ+i\sin 20^\circ)]$
**4.** $[5(\cos 2\pi/3+i\sin 2\pi/3)][2(\cos\pi/3+i\sin\pi/3)]$
**5.** $[4(\cos\{-70^\circ\}+i\sin\{-70^\circ\})][2(\cos 15^\circ+i\sin 15^\circ)]$

Using the symbol $r$ cis $\theta$ to represent the complex number $r$ (cos $\theta$ + $i$ sin $\theta$), the product law may be stated as

$$[r_1 \text{ cis } \theta_1][r_2 \text{ cis } \theta_2] = r_1 r_2 \text{ cis } (\theta_1 + \theta_2)$$

Write each of the following products in the simplest form.

**6.** $(3 \text{ cis } 25°)(2 \text{ cis } 35°)$

**7.** $(2 \text{ cis } \pi/2)(4 \text{ cis } \tfrac{3}{4}\pi)$

**8.** $(4 \text{ cis}(-\pi))(3 \text{ cis } \pi/2)$

**9.** $(5 \text{ cis } 200°)(2 \text{ cis } 25°)$

In Exercises **10** to **21** use De Moivre's formula to aid you in expressing each of the complex numbers in the form $a + bi$.

**10.** $[2(\cos 10° + i \sin 10°)]^3$

**11.** $(1 - i)^2$

**12.** $(3 \text{ cis } \pi/3)^4$

**13.** $(1 + i)^5$

**14.** $(2 \text{ cis } -18°)^5$

**15.** $(\sqrt{3} + i)^3$

**16.** $(3 \text{ cis } 100°)^3$

**17.** $(-i)^5$

**18.** $(-\sqrt{3} + i)^2$

**19.** $(3)^4$

**20.** $(2 - 2i)^4$

**21.** $(2 \text{ cis } 90°)^3$

**22.** Express cos $4\theta$ and sin $4\theta$ in terms of cos $\theta$ and sin $\theta$ by using the method of Example 2 in (9.3).

**23.** Express cos $2\theta$ and sin $2\theta$ in terms of sin $\theta$ and cos $\theta$ by using the same method.

---

## 9.4 THE QUOTIENT OF TWO COMPLEX NUMBERS IN POLAR FORM

Given two complex numbers,

$$z_1 = r_1 (\cos \theta_1 + i \sin \theta_1)$$

and

$$z_2 = r_2 (\cos \theta_2 + i \sin \theta_2)$$

then

$$\frac{z_1}{z_2} = \frac{r_1 (\cos \theta_1 + i \sin \theta_1)}{r_2 (\cos \theta_2 + i \sin \theta_2)}, \quad r_2 \neq 0$$

If the numerator and denominator of the right-hand member are multiplied by cos $\theta_2$ − $i$ sin $\theta_2$, the equation reduces to

$$\begin{aligned} \frac{z_1}{z_2} &= \frac{r_1}{r_2} \frac{\cos \theta_1 \cos \theta_2 + \sin \theta_1 \sin \theta_2 + i (\sin \theta_1 \cos \theta_2 - \cos \theta_1 \sin \theta_2)}{\cos^2 \theta_2 + \sin^2 \theta_2} \\ &= \frac{r_1}{r_2} [\cos (\theta_1 - \theta_2) + i \sin (\theta_1 - \theta_2)], \quad r_2 \neq 0 \end{aligned}$$

In abbreviated form,

$$\frac{r_1 \text{ cis } \theta_1}{r_2 \text{ cis } \theta_2} = \frac{r_1}{r_2} \cos (\theta_1 - \theta_2)$$

## EXERCISES

Use the quotient rule of two complex numbers in polar form to simplify Exercises **1** to **6**.

1. $\dfrac{4(\cos 42° + i \sin 42°)}{2(\cos 12° + i \sin 12°)}$
2. $\dfrac{12[\cos (3\pi/4) + i \sin (3\pi/4)]}{6(\cos \pi/2 + i \sin \pi/2)}$
3. $\dfrac{5(\cos 4\pi + i \sin 4\pi)}{3(\cos 3\pi + i \sin 3\pi)}$
4. $\dfrac{14 \text{ cis } 210°}{7 \text{ cis } 120°}$
5. $\dfrac{4 \text{ cis } \pi}{3 \text{ cis } 2\pi}$
6. $\dfrac{\text{cis } 45°}{2 \text{ cis } 15°}$
7. Prove that, if $r \neq 0$,

$$[r (\cos \theta + i \sin \theta)]^{-1} = \frac{1}{r} [\cos (-\theta) + i \sin (-\theta)]$$

that is to say, *verify De Moivre's formula for* $n = -1$. [*Hint:* Note that

$$[r (\text{cis } \theta)]^{-1} = \frac{1}{r (\text{cis } \theta)} = \frac{\text{cis } 0}{r(\text{cis } \theta)}$$

and use the quotient rule.]

8. Prove De Moivre's formula for $n$ any negative integer. (*Hint:* Take the $n$th power of both members of the identity proved in Ex. 7.)
9. Simplify $[2 (\cos 10° + i \sin 10°)]^{-3}$
10. If $z_1 = x_1 + iy_1$ and $z_2 = x_2 + iy_2$, show geometrically that the smallest positive amplitude of $(z_1 - z_2)$ is the angle between the right-hand horizontal direction and the line $z_2z_1$.
11. Show that the amplitude of $\dfrac{z_1 - z_2}{z_3 - z_2}$ is $\angle Z_3Z_2Z_1$.
12. *A mettle tester:* If $Z_1$, $Z_2$, $Z_3$, and $Z_4$ are distinct points on a circle, show that

$$\frac{\dfrac{z_1 - z_2}{z_4 - z_2}}{\dfrac{z_1 - z_3}{z_4 - z_3}} \text{ is a real number.}$$

*Hint:* Use the fact that on the circle, $\angle Z_1Z_2Z_4 = \angle Z_1Z_3Z_4$.

## 9.5 ROOTS OF COMPLEX NUMBERS

We next discuss methods of evaluating $\sqrt{3 - 4i}$, $\sqrt[3]{4 - 2i}$, $\sqrt[7]{3 + 7i}$, and similar expressions, or equivalently of solving

$$x^2 = 3 - 4i, \quad x^3 = 4 - 2i, \quad x^7 = 3 + 7i, \quad \text{and so forth}$$

We shall find that further definition must be given to these radical forms if ambiguity is to be avoided.

In seeking for an $n$th root of a complex number $a + bi$, we first express the number in the general polar form

$$a + bi = r\,[\cos (\theta + k \cdot 2\pi) + i \sin (\theta + k \cdot 2\pi)]$$

or

$$a + bi = r \text{ cis } (\theta + 2k\pi), \quad k = 0, \pm 1, \pm 2, \ldots$$

Now, if $z$ is such a complex number that $z^n = a + bi$, then $z$ is an $n$th root of $a + bi$. Let $z = s\,(\cos \varphi + i \sin \varphi) = s \text{ cis } \varphi$. Then, since it is assumed that $z^n = a + bi$,

$$(s \text{ cis } \varphi)^n = r\,[\text{cis } (\theta + 2k\pi)]$$

Hence, by De Moivre's formula,

$$s^n \text{ cis } n\varphi = r\,[\text{cis } (\theta + 2k\pi)]$$

This equation is satisfied if, and only if,

$$s^n = r \quad \text{and} \quad n\varphi = \theta + 2k\pi$$

that is, if, and only if,

$$s = \sqrt[n]{r} \quad \text{and} \quad \varphi = \frac{\theta}{n} + \frac{2k\pi}{n}, \quad k = 0, \pm 1, \pm 2, \ldots$$

where $\sqrt[n]{r}$ is the positive, real $n$th root of the positive number $r$. The $n$ different choices of $\varphi$ obtained by using $k = 0, 1, 2, \ldots, n - 1$, respectively, determined $n$ distinct $n$th roots of $a + bi$. It will be found that other integral values of $k$ give values of $\varphi$ co-terminal with those found by the above choices of $k$. We therefore conclude that there are $n$ distinct $n$th roots of $a + bi = r \text{ cis } \theta$ and that they are given by the formula

$$z = \sqrt[n]{r} \text{ cis } \left(\frac{\theta}{n} + \frac{2k\pi}{n}\right), \quad k = 0, 1, 2, 3, \ldots, n - 1 \qquad (9.2)$$

**EXAMPLE 1**

Find the three cube roots of $8i$.

**Solution:** In polar form,

$$8i = 8 \text{ cis } 90^\circ$$

Hence, by (9.2), the cube roots of $8i$ are given by

$$z = \sqrt[3]{8} \text{ cis} \left( \frac{90°}{3} + \frac{k \cdot 360°}{3} \right), \quad k = 0, 1, 2$$

If $k = 0, 1, 2$, determine the roots $z_1, z_2, z_3$, respectively; then

$$\begin{aligned} z_1 &= 2 \text{ cis } 30° \\ &= 2\left(\frac{\sqrt{3}}{2} + \frac{i}{2}\right) \\ &= \sqrt{3} + i \\ z_2 &= 2 \text{ cis } 150° \\ &= 2\left(-\frac{\sqrt{3}}{2} + \frac{i}{2}\right) \\ &= -\sqrt{3} + i \end{aligned}$$

and

$$\begin{aligned} z_3 &= 2 \text{ cis } 270° \\ &= 2(0 - i) \\ &= -2i \end{aligned}$$

The cube roots of $8i$ are shown graphically in Figure 9.5.

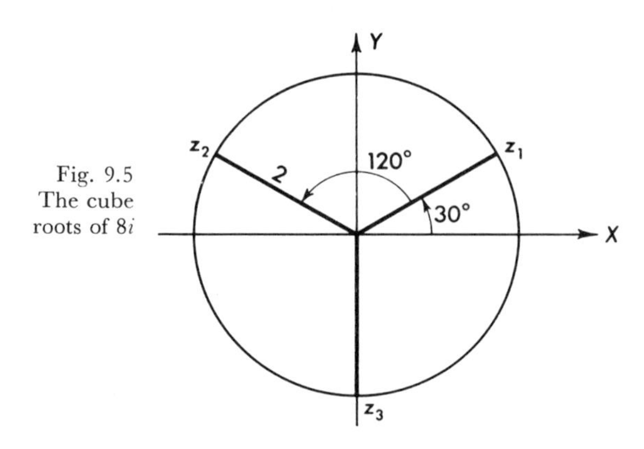


Fig. 9.5
The cube
roots of $8i$

**EXAMPLE 2**

Solve the equation $z^5 + 4 + 4i = 0$.

**Solution:** We have

$$\begin{aligned} z^5 &= -4 - 4i \\ &= \sqrt{32} \text{ cis } 225° \end{aligned}$$

Hence

$$z = \sqrt[10]{32} \text{ cis} \left( \frac{225}{5} + \frac{360k}{5} \right)^{\circ}$$

or

$$z = \sqrt{2} \text{ cis } (45 + 72k)°, \quad k = 0, 1, 2, 3, 4$$

For:

$$\begin{aligned} k &= 0, & z_1 &= \sqrt{2} \text{ cis } 45° \\ k &= 1, & z_2 &= \sqrt{2} \text{ cis } 117° \\ k &= 2, & z_3 &= \sqrt{2} \text{ cis } 189° \\ k &= 3, & z_4 &= \sqrt{2} \text{ cis } 261° \\ k &= 4, & z_5 &= \sqrt{2} \text{ cis } 333° \end{aligned}$$

If desired, the solutions may be written in the form $a + bi$, $z_1$ exactly, in this example, and $z_2$, $z_3$, $z_4$, and $z_5$ with approximate values from the table of natural functions. Thus

$$\begin{aligned} z_1 &= 1 + i \\ z_2 &\approx \sqrt{2}\,(-0.454 + 0.891i) \\ &\approx -0.645 + 1.260i \end{aligned}$$

and so on. The five roots of the given equation are shown graphically in Figure 9.6.

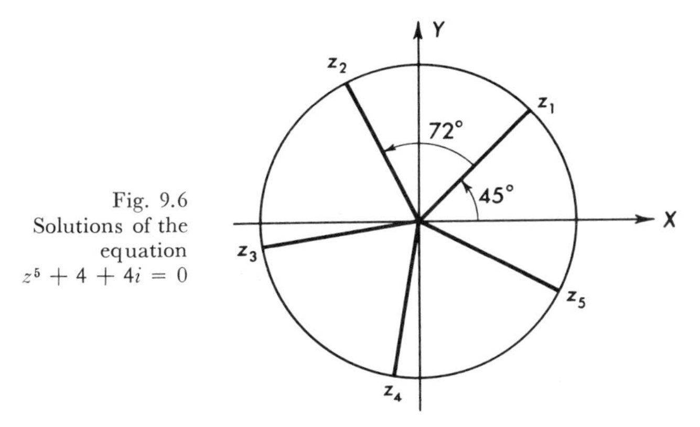


Fig. 9.6 Solutions of the equation $z^5 + 4 + 4i = 0$

Observe that graphically the $n$ $n$th roots of a complex number $r$ cis $\theta$ are points which lie on the circle with center at the origin and with radius $\sqrt[n]{r}$. These points divide the circle into $n$ equal arcs. *We now define the symbol* $\sqrt[n]{a + bi}$ *to be* (a) the real $n$th root of smallest non-negative amplitude, if there is a real root; (b) the root of smallest non-negative amplitude, if there is no real root. It is in keeping with this definition that $\sqrt{4}$ is 2, not $-2$ or $+2$, and that $\sqrt[3]{-8} = -2$. You will find it enlightening to determine these roots by the above described method.

## *EXERCISES*

1. Show that the cube roots of 1 are 1, $-\frac{1}{2}+\frac{i\sqrt{3}}{2}$, and $-\frac{1}{2}-\frac{i\sqrt{3}}{2}$. Display these roots graphically and observe that $\sqrt[3]{1} = 1$, not $-\frac{1}{2}+\frac{i\sqrt{3}}{2}$ or $-\frac{1}{2}-\frac{i\sqrt{3}}{2}$, in keeping with the final discussion above.
2. Find the square roots of $i$, using the methods of this section. (Compare with Exercise 21, Section 9.1.)
3. Show that the cube roots of any real number $b$ are the products of $\sqrt[3]{b}$ and the respective cube roots of 1.
4. Use Exercises 1 and 3 to find the cube roots of

   (**a**) 8 (**b**) $-27$
5. Find the square roots of $-1 + i$.
6. Solve the equation $z^3 - 4\sqrt{3} + 4i = 0$.
7. Solve the equation $z^4 + 1 = 0$.
8. Show that, if $z_i$ and $z_j$ are $n$th roots of 1, then $z_i z_j$ is also an $n$th root of 1.
9. Show that, if $\alpha = \text{cis}\left(\frac{360}{n}\right)^\circ$, an $n$th root of 1, then $\alpha^2, \alpha^3, \alpha^4, \ldots, \alpha^n$ are the remaining distinct $n$th roots of 1.
10. Explain why it is that, graphically, (**a**) all of the $n$th roots of a given complex number lie on a circle, and (**b**) they divide the circle into $n$ equal arcs.

## *MISCELLANEOUS EXERCISES/CHAPTER 9*

1. Find the modulus and smallest non-negative amplitude of the following.

   (**a**) $1 - i$ (**b**) $2 + i$ (**c**) $2i$ (**d**) 4
2. Simplify

   (**a**) $(2 + i)(2 - i) - (3 + 4i)(1 - i)$ (**b**) $(i)^{25}$ (**c**) $(i)^{30}$
3. Simplify

   (**a**) $\frac{2}{(2-2i)}$ (**b**) $\frac{(1-i)}{2+2i}$ (**c**) $\frac{(3+4i)}{i}$
4. If $|z|$ is the absolute value, or modulus, of $z$, and if $\bar{z}$ is the conjugate of $z$ (see Section 9.1), verify the following.

(a) $|z| = \sqrt{z\bar{z}}$.

(b) If $z = r \text{ cis } \theta$, then $\bar{z} = r \text{ cis } (-\theta)$.

(c) Show the points $z$, $\bar{z}$, and $|z|$ in the complex plane for several choices of $z$.

**5.** Find graphically the value of $(-2 + 3i) + (1 + 2i)$ and check algebraically.

**6.** Find graphically the value of $(2 - 3i) - (1 - 2i)$ and check algebraically.

**7.** Let $F$ and $G$ be two forces in a vertical plane, and let $F_x$ and $F_y$ be the respective horizontal and vertical components of $F$. If $F_x = 30$ lb., $F_y = 50$ lb., $G_x = -20$ lb., and $G_y = 40$ lb., represent the forces $F$ and $G$ by complex numbers. Find their resultant (sum) and reduce it to the polar form. From this, give the magnitude and direction of the resultant force.

In Exercises **8** through **15**, reduce to the simplest form.

**8.** $\dfrac{3(\cos 25° + i \sin 25°)(\cos 75° + i \sin 75°)}{4(\cos 10° + i \sin 10°)(\cos 30° + i \sin 30°)}$

**9.** $\dfrac{2(\cos C + i \sin C)(\cos D + i \sin D)}{\cos (-C) + i \sin (-C)}$

**10.** $(4 \text{ cis } 35\frac{1}{2}°)(3 \text{ cis } 64\frac{1}{2}°)$

**11.** $(2 \text{ cis } 15°)^4$

**12.** $(1 - i)^5$

**13.** $(\sqrt{3} + i)^4$

**14.** $(-i)^{12}$

**15.** $(2i)^3$

**16.** Find the 5th roots of $2 + 2\sqrt{3}\, i$.

**17.** Find the 4th roots of $-i$.

**18.** Express $(1 - \sqrt{3}\, i)^5$ as a complex number.

**19.** Simplify $\dfrac{(\cos 4\theta + i \sin 4\theta)(\cos 2\theta + i \sin 2\theta)}{\cos 5\theta - i \sin 5\theta}$.

(Be careful).

**20.** Simplify $\dfrac{(\cos 3\theta + i \sin 3\theta)}{(\cos 2\theta - i \sin 2\theta)(\cos 5\theta + i \sin 5\theta)}$.

(Be careful).

**21.** *A mettle tester.* Prove that De Moivre's formula,

$$(\text{cis } \theta)^n = \text{cis } n\theta,$$

is valid for $n = \dfrac{p}{q}$, where $p$ and $q$ are integers with $q > 0$. [*Hint:* Assume that $z^{p/q}$ means $(t)^p$, where $t = \sqrt[q]{z}$ as defined in the last paragraph of Section 9.5. Note that $p$ may be negative.]

# APPENDIX

The Appendix contains a discussion of two topics: I, *Accuracy of computed results*, and II, *Logarithms*. Exercises are provided so that the material may be used either as supplementary class work or as self-directed review by the student.

## I. ACCURACY OF COMPUTED RESULTS

The most important consideration connected with accuracy of answers is the question of the accuracy of the numbers used to obtain the answer. Some of the numbers will be *exact*, as for example the 2 in the formula $c = 2\pi r$, others will be *approximate*, as are all numbers obtained by measurement and most of the numbers obtained from tables. (Exceptions are such numbers as $\sin 30° = 0.5$, $\log 100 = 2$, and $\sqrt{16} = 4$, which are exact.) Irrational numbers such as $\pi$, $\sqrt{2}$, and $\sqrt[3]{5}$ are often represented by approximate decimal expressions.

One way of expressing the degree of accuracy of an approximate number is to state the number of significant figures (or digits) to which it is accurate. Thus, if the number $\pi$ is equal to 3.14159265358979 . . . , where the dots mean that the digits may be continued indefinitely, it is customary to say that 3.14, 3.142, 3.1416, and 3.141593 are approximations of $\pi$ correct to 3, 4, 5, and 7 significant figures, respectively. *Significant figures* are digits that give meaning to a number regardless of the location of the decimal point. The digits 1, 2, 3, 4, 5, 6, 7, 8, 9 are always significant when present in a number. Zero is not significant if it occurs to the left of all the other digits in a number, as in 0.45 or 0.00045; it is significant if it occurs between other digits, as in 205 and 6.007; its significance is in doubt if it occurs to the right of the other digits as in 40 or 3300. If the approximate number 40 actually represents a certain magnitude more exactly than 39 or 41 does, then the figure 0 is significant, but, if it merely represents a number closer to 40 than to 30 or 50, the 0 is not significant. It is proper to be specific in such cases with a statement. For example, if 3300 is a better approximation to a certain measurement than 3290 or 3310, with no other information as to its accuracy, it is said to be correct to 3 significant figures, or correct to the nearest 10. It will be assumed that a zero is significant if it is written to the right of the decimal point and of all other digits. Thus 7.50 is considered accurate to three significant figures, and 53.00 to four significant figures.

As a chain is no stronger than its weakest link, so a numerical result obtained by multiplications and divisions with approximate numbers will, as a rule, contain no more significant figures than that one of the given numbers having the least number of significant figures. This is a safe working rule. As an illustration, suppose that the radius of a circle is 22 in. to the nearest inch, and that it is desired to know the area of the circle from the formula $\pi r^2$. If the value of $\pi$ is taken as the approximate number 3.1416, the value of $\pi r^2$ as found from the formula is 1530.5344 sq in., an obvious absurdity—even worse, a delusion of accuracy. Actually, all that can be said is that the area is 1500 sq in., correct to two significant figures. On the other hand, if the radius is 22.000, correct to five significant figures, the answer is 1530.5 sq in., since both component numbers were *correct* to five significant figures.

A simple rule for the addition or subtraction of approximate numbers is that the result contains no more accuracy than the least *decimal place accuracy* of the given terms. For instance, the sum $3.14 + 15.812 + 0.5933$ should be rounded off to *two* decimal places. Example 2 helps to show why this is so.

In rounding off numbers to a prescribed number of significant digits, the choice of whether to leave the last retained digit unchanged or to increase it by one is too familiar to the reader to need repetition except in the case that the discarded digits amount to exactly half a unit of the desired accuracy. It is considered good practice by computers to make the last retained digit an *even* one in such cases. Thus the numbers 0.68645, 0.42375, and 3,426,500 become 0.6864, 0.4238, and 3,426,000, respectively, when rounded off to accuracy of four significant figures. Judicious rounding off of numbers will save time without sacrifice of accuracy if done *before* computation is begun. If in the illustration of the area of a circle with radius 22 in., the value of $\pi$ had been rounded off to 3.14, the area could have been computed more rapidly and with the same accuracy as when 3.1416 was used.

The accuracy with which angles are determined by means of tables of their trigonometric functions depends somewhat upon the size of the angle. However, as a general rule, functions expressed with 3, 4, 5, or 6 significant figures correspond in accuracy of the angle to the nearest tenth of a degree, a minute, tenth of a minute, or a second, respectively.

### EXAMPLE 1

Find the product of the approximate numbers 4384 and 573, showing why the product must be rounded off to three significant digits.

**Solution:** As an approximate number, the symbol 4384 represents some number between 4383.5 and 4384.5. Likewise the approximate number 573 is between 572.5 and 573.5. Hence the product

$$(4384)(573) = 2{,}512{,}032$$

represents some number between the product

$$(4383.5)(572.5) = 2{,}509{,}553.75$$

and the product

$$(4384.5)(573.5) = 2{,}514{,}510.75$$

These three products agree to three significant digits only. Thus the product is written

$$2{,}510{,}000$$

### EXAMPLE 2

Find the sum of the approximate numbers 23.17, 6.132, 157.666, and 493.4, showing why it contains no more accuracy than the least decimal place accuracy of the given terms.

**Solution:** Below, the given sum is exhibited as lying between the sum of the smallest possible values of the given approximate numbers and the sum of their largest possible values.

| | | |
|---|---|---|
| 23.165 | 23.17 | 23.175 |
| 6.1315 | 6.132 | 6.1325 |
| 157.6655 | 157.666 | 157.6665 |
| 493.35 | 493.4 | 493.45 |
| 680.3120 | 680.368 | 680.4240 |

Examination of these sums shows that 680.4 is the most optimistic estimate of the actual value.

## *EXERCISES*

**1.** State the number of significant digits in each of the following numbers.

(**a**) 0.0563 (**b**) 0.507
(**c**) 5.070 (**d**) 60,308
(**e**) 3.14159

*Ans.* (**a**) three (**c**) four (**e**) six

**2.** Round off each of the following approximate numbers to three significant digits

(**a**) 31.582 (**b**) 0.005123

(**c**) 8.1253 (**d**) 8.125

(**e**) 62,458,702 (**f**) 0.9735

*Ans.* (**a**) 31.6 (**c**) 8.13 (**f**) 0.974

**3.** One way to indicate the number of significant digits in an approximate number is to write it in *standard form*, that is, to write it in the form $n(10^k)$, where $1 \leqq n < 10$ and $k$ is an integer. In this manner, $n$ carries the significant digits, while $10^k$ locates the decimal point. Examples of numbers expressed in standard form are

(**a**) 653,400,000 (correct to four digits) $= 6.534(10^8)$

(**b**) 2,920,000,000 (correct to *four* digits) $= 2.920(10^9)$

(**c**) $0.0000072 = 7.2(10^{-6})$. $\left(\text{Remember that } 10^{-6} = \dfrac{1}{10^6}\right)$

In similar fashion express each of the following numbers in standard form:

(**d**) 874 *Ans.* $8.74(10^2)$

(**e**) 283,000, correct to three digits

(**f**) 56.23 *Ans.* $5.623(10^1)$

(**g**) 283,000, correct to four digits

(**h**) 0.00512 *Ans.* $5.12(10^{-3})$

(**i**) 0.3912

(**j**) 0.00003 *Ans.* $3(10^{-5})$

(**k**) 3,000,000,000, correct to three digits

**4.** State the number of significant figures in the result of each of the following computations. The actual computation is not required. Assume that the given numbers are approximate.

(**a**) (3.1416)(51)(124) *Ans.* Two

(**b**) (6)(107)

(**c**) $382 \div 566$ *Ans.* Three

(**d**) $\dfrac{(0.6814)(0.0307)}{45.87}$

(**e**) the sum

$$\begin{array}{r} 86.344 \\ 292.32 \\ 6.5036 \\ \hline \end{array}$$

*Ans.* Five

(**f**) $2.082 + 989.4 - 25.17$

---

## II. COMMON LOGARITHMS

The numbered paragraphs that follow give in brief compass the essentials of common logarithms. The sentences in italics, separated from their respective explanations, serve as an outline of the development. It is assumed that the student is familiar with the symbol

$$a^x$$

and its meaning if $a$ is any positive number and $x$ is any real number. The following definitions from algebra will therefore be assumed:

$$a^{-x} = \frac{1}{a^x}$$

$$a^0 = 1$$

$$a^{p/q} = \sqrt[q]{a^p}, \quad \text{if } p \text{ and } q \text{ are integers,} \quad q \neq 0.*$$

We also assume the following laws of exponents,

$$(a^x)(a^y) = a^{x+y}$$

$$\frac{a^x}{a^y} = a^{x-y}$$

$$(a^x)^y = a^{xy}$$

Before confining our attention to the subject of common, or base-ten, logarithms, let us consider a general definition of logarithms. If $a$ is any positive number different from 1, and if $N$ is any positive number, it is shown in more advanced courses that the equation

$$a^x = N$$

is satisfied by exactly one real value of $x$. This root of the equation is called the logarithm of $N$ to the base $a$. In abbreviated form,

$$x = \log_a N$$

The equations $a^x = N$ and $x = \log_a N$ are equivalent statements of the same relationship. Substitution of the value of $x$ in the second into the first equation results in the equation

$$a^{\log_a N} = N$$

This equation forms the definition of $\log_a N$. Thus we have a logarithm function $l$: $P \rightarrow R$, whose domain is the set of positive real numbers $P$ and whose range is the set of real numbers $R$, such that, if $N \in P$ then $\log_a N \in R$.

**EXAMPLE 1**

Find $\log_2 32$.

* As indicated above, it is assumed that a $> 0$. It may be added that $0^x = 0$ if $x \neq 0$, and that $a^{p/q} = \sqrt[q]{a^p}$ if $a < 0$ and $p$ is even.

**Solution:** By definition,

$$2^{\log_2 32} = 32$$

But

$$2^5 = 32$$

Hence

$$\log_2 32 = 5$$

**EXAMPLE 2**

If $\log_8 N = 2/3$, find $N$.
**Solution:** By definition,

$$N = 8^{\log_8 N}$$

Hence

$$N = 8^{2/3} = \sqrt[3]{8^2} = \sqrt[3]{64} = 4$$

## *EXERCISES*

**1.** Express each of the following exponential equations in logarithmic form.
(**a**) $e^r = k$ (**b**) $5^x = 75$
(**c**) $10^x = 1000$ (**d**) $b^3 = M$

**2.** Express each of the following logarithmic equations in exponential form.
(**a**) $y = \log_b T$ (**b**) $\log_3 K = x$
(**c**) $\log_9 27 = 3/2$ (**d**) $\log_{10} 0.01 = -2$

**3.** Find the value of the unknown in each of the following equations.
(**a**) $4^x = 16$ (**b**) $5^x = 125$ *Ans.* $x = 3$
(**c**) $5^x = 0.2$ (**d**) $10^x = 0.1$ *Ans.* $x = -1$
(**e**) $4^x = 8$ (**f**) $16^x = 64$ *Ans.* $x = 3/2$.
(**g**) $8^x = 1/4$ (**h**) $10^x = 1$ *Ans.* $x = 0$
(**i**) $x = \log_3 81$ (**j**) $x = \log_{25} 625$ *Ans.* $x = 2$
(**k**) $x = \log_{10} 1000$ (**l**) $x = \log_{100} 1000$ *Ans.* $x = 3/2$
(**m**) $x = \log_{1/4} 16$ (**n**) $x = \log_4 1/16$ *Ans.* $x = -2$
(**o**) $x = \log_{10} 0.01$ (**p**) $x = \log_{27} 9$ *Ans.* $x = 2/3$
(**q**) $\log_x 36 = 2$ (**r**) $\log_x 64 = 6$ *Ans.* $x = 2$
(**s**) $\log_x 5 = 0.5$ (**t**) $\log_x 27 = 1.5$ *Ans.* $x = 9$
(**u**) $\log_x 10{,}000 = 4$ (**v**) $\log_x 0.1 = -1$ *Ans.* $x = 10$

**4.** Simplify
(**a**) $z^{\log_z R}$ (**b**) $e^{\log_e x}$ *Ans.* $x$
(**c**) $\log_y y$ (**d**) $\log (a^{\log_a x})$ *Ans.* $\log x$
(**e**) $\log_a 1$ (**f**) $\log_k k$ *Ans.* 1

For purposes of computation, a system of logarithms using base 10 is convenient. This is due to the underlying use of the base 10 in the decimal system of writing numbers. Logarithms of numbers to the base 10 are called *common logarithms*, or simply *logarithms*. Hence the word logarithm will mean *logarithm to the base 10* unless another base is explicitly given. Consider now the following outlined discussion of the meaning and use of common logarithms:

**1.** *Every positive number can be expressed as a power of ten.* This is to say that, for every positive number $N$, there exists a real number $x$ such that

$$N = 10^x$$

This follows from the more general statement made above that, if $N > 0$ and $a > 0$ but $a \neq 1$, then the equation

$$N = a^x$$

is satisfied by exactly one real value of $x$. The proof is beyond the scope of this book, but the statement will seem more reasonable by observing the following illustrations of various numbers expressed as powers of 10:

$$\begin{aligned}
1000 &= 10^3 \\
10 &= 10^1 \\
1 &= 10^0 \\
0.01 &= \frac{1}{10^2} = 10^{-2} \\
3.162 &\approx \sqrt{10} \approx 10^{0.5000} \\
2.154 &\approx \sqrt[3]{10} \approx 10^{0.3333} \\
31.62 &\approx \sqrt{10^3} = 10^{3/2} \approx 10^{1.5000} \\
0.3162 &\approx \frac{1}{\sqrt{10}} \approx 10^{-0.5000}
\end{aligned}$$

**2.** *If $N = 10^x$, the exponent $x$ is called the common logarithm, or simply the logarithm, of the number $N$.* This is equivalent to the equation

$$N = 10^{\log N}$$

Thus the logarithm of 100 is 2 because $10^2 = 100$; the logarithm of 2.154 is approximately 0.3333 because $10^{0.3333} \approx 2.154$; $\log 0.3162 \approx 0.5$ because $10^{-0.5} \approx 0.3162$; $\log 1 = 0$ because $10^0 = 1$; and so forth.

**3.** *The logarithms of numbers which are integral (whole number) powers of ten are obtained by inspection.* This statement has already been put into effect in the preceding discussion and is made here for the sake of emphasis. The student can readily determine such results as

$$\begin{aligned}
\log 100{,}000 &= 5 \\
\log 0.001 &= -3
\end{aligned}$$

and

$$\log 100 = 2$$

**4.** *The logarithms of numbers of three digits from 1.00 to 9.99, inclusive, are given to four decimal places in Table II.* To find the logarithm of such a number in the table, locate the first two digits in the column headed by $n$, hold their position, and proceed horizontally to the right until the column headed by the third digit of the given number is reached. The number thus located, with a decimal point placed at its left, is the logarithm of the given number. For example, to find log 5.62, look down the column under $N$ until 56 is located (all decimal points are omitted in the table), then move across the table until the column headed by 2 is reached. The corresponding entry, 0.7497, is the logarithm of 5.62. As other examples, the student should verify that log 1.03 = 0.0128 and that log 7.48 = 0.8739.

**5.** *Logarithms can be used to simplify computations involving the operations of multiplication, division, and finding powers and roots.* Consider a few illustrative examples.

**EXAMPLE 1**

Find

$$\frac{(3.23)(2.99)}{2.21}$$

**Solution:** From the tables,

$$\begin{aligned} \log 3.23 &= 0.5092, \quad \therefore\ 3.23 = 10^{0.5092} \\ \log 2.99 &= 0.4757, \quad \therefore\ 2.99 = 10^{0.4757} \\ \log 2.21 &= 0.3444, \quad \therefore\ 2.21 = 10^{0.3444} \end{aligned}$$

Hence, using the rules that $(a^x)(a^y) = a^{x+y}$ and that $\frac{a^x}{a^y} = a^{x-y}$,

$$\begin{aligned} \frac{(3.23)(2.99)}{2.21} &= \frac{(10^{0.5092})(10^{0.4757})}{10^{0.3444}} \\ &= 10^{0.5092+0.4757-0.3444} \\ &= 10^{0.6405} \end{aligned}$$

Reference to the tables will show that the number whose logarithm is 0.6405 is 4.37. Thus $\frac{(3.23)(2.99)}{2.21} \approx 4.37$.

(*Note:* The reader may not agree that this method "simplifies" the operations, but he certainly must agree that the operations of multiplication and division have been reduced to the *simpler* operations of addition and subtraction. Later, improved methods of manipulating the logarithms will be shown.)

### EXAMPLE 2

Find $\sqrt[3]{9}$.
**Solution:** After the manner of Exercise 1, $\log 9 = 0.9542$, so that $9 = 10^{0.9542}$. Hence

$$\sqrt[3]{9} = \sqrt[3]{10^{0.9542}} = (10)^{\frac{0.9542}{3}} = 10^{0.3181}$$

From the table, 0.3181 is the logarithm of 2.08. Hence

$$\sqrt[3]{9} \approx 2.08$$

## *EXERCISES*

Apply the method of the above examples to compute each of the following, correct to the nearest third digit.

1. $(4.33)(1.94)$ *Ans.* 8.40
2. $6.5\sqrt{2.88}$
3. $\dfrac{8.26}{2.67}$ *Ans.* 3.09
4. $\dfrac{(7.57)(4.9)}{5.18}$
5. $(1.06)^{20}$ *Ans.* 3.21
6. $(7.36)^{2/3}$
7. $\sqrt{7.51}$ *Ans.* 2.74
8. $\sqrt{(6.41)(3.29)}$
9. $\sqrt[3]{8.92}$ *Ans.* 2.07
10. $\dfrac{9.37}{(2.4)(3.25)}$

**6.** *By a process of interpolation, the logarithm of any four-digit number from 1.000 to 9.999 can be determined approximately from the table. Conversely, given the logarithm of any such number, the number can be determined correct to four digits.* These statements can best be explained by the use of examples.

### EXAMPLE 1

Find log 1.954.
**Solution:** The value of log 1.954 obviously is between that of log 1.950 and of log 1.960. In fact, it would seem to be four-tenths "of the way" between them. (Experience shows that this assumption of proportionality is accurate enough to permit the use of the tables in finding the logarithms of numbers within the range stated above.) Upon this assumption,

$$\begin{aligned}\log 1.954 &= \log 1.950 + 0.4(\log 1.960 - \log 1.950)\\ &= 0.2900 + 0.4(0.2923 - 0.2900)\\ &= 0.2900 + 0.4(.0023)\\ &= 0.2909\end{aligned}$$

In actual practice, since all the logarithms are to ten-thousandths, it is customary to make the subtraction mentally without copying from the table, obtaining 23 (ten-thousandths,) multiply by 0.4, obtaining 9 (ten-thousandths), and then add to the first logarithm.

**EXAMPLE 2**

Find the number whose logarithm is 0.2420.

**Solution:** Examination of the table shows that 0.2420 lies between the logarithms of 1.740 and 1.750. Thus, if $\log n = 0.2420$,

$$\log 1.740 = 0.2405$$
$$\log n = 0.2420$$
$$\log 1.750 = 0.2430$$

Assuming proportionality, it follows that $n$ is approximately $15/25 = 0.8$ "of the way" between 1.740 and 1.750. Hence

$$n \approx 1.748$$

## *EXERCISES*

Find the logarithms of the following numbers.

**1.** 3.655. *Ans.* 0.5629.
**2.** 8.202.
**3.** 8.912. *Ans.* 0.9500.
**4.** 3.645.
**5.** 2.576. *Ans.* 0.4109.
**6.** 1.144.
**7.** 3.142. *Ans.* 0.4972.
**8.** 5.338.
**9.** 3.1416. *Ans.* 0.4972.
**10.** 5.337831.
**11.** 4.004. *Ans.* 0.6025.
**12.** 8.999.

Find correct to four digits the numbers whose logarithms are

**13.** 0.4806. *Ans.* 3.024.
**14.** 0.6175.
**15.** 0.6388. *Ans.* 4.353.
**16.** 0.3434.
**17.** 0.8303. *Ans.* 6.766.
**18.** 0.3314.
**19.** 0.4400. *Ans.* 2.754.
**20.** 0.5701.
**21.** 0.6605. *Ans.* 4.576.
**22.** 0.6301.
**23.** 0.8245. *Ans.* 6.676.
**24.** 0.3844.

**7.** *Since logarithms are exponents, indicating powers of ten, they obey the laws of exponents so that laws of logarithms are:*

(I) $$\log (A \cdot B) = \log A + \log B,$$

(II) $$\log \left(\frac{A}{B}\right) = \log A - \log B,$$

(III) $$\log (A)^p = p \log A,$$

(IV) $$\log \sqrt[r]{A} = \frac{1}{r} \log A.$$

**PROOF OF (I)**

In the same way that $10^2 \times 10^3 = 10^{2+3} = 10^5$, it is seen that

$$A \cdot B = 10^{\log A} \times 10^{\log B} = 10^{\log A + \log B},$$

which is equivalent to (I) above. The other three laws can be verified similarly.

**EXAMPLE**

Use the laws of logarithms to express the logarithm of $N$ if

(**a**) $N = x^3y\sqrt{4 + x^2}$
(**b**) $N = (3.14)(10^2)$

**Solutions:**

(**a**) Combining the logarithm laws for products (I), for powers (III), and for roots (IV), $\log N$ can be written

$$\log N = 3 \log x + \log y + \tfrac{1}{2} \log (4 + x^2)$$

(**b**) If the product law is applied, log 3.14 determined from the table, and $\log 10^2$ found by inspection, then

$$\begin{aligned} \log N &= \log 3.14 + \log (10^2) \\ &= 0.4969 + 2 \\ &= 2.4969 \end{aligned}$$

## *EXERCISES*

Express the logarithm of both members of the equation in each of the following and find the value of the logarithm in the numerical cases.

**1.** $K = \sqrt{s(s - a)(s - b)(s - c)}$
*Ans.* $\log K = \tfrac{1}{2}[\log s + \log (s - a) + \log (s - b) + \log (s - c)]$

**2.** $x = \sqrt{\dfrac{(s - b)(s - c)}{s(s - a)}}$.

**3.** $V = \tfrac{4}{3}\pi r^3$ *Ans.* $\log V = \log 4 + \log \pi + 3 \log r - \log 3$

**4.** $l = ar^{k-1}$

**5.** $n = ke^x$ *Ans.* $\log n = \log k + x \log e$

**6.** $S = P(1 + i)^n$

**7.** $N = 4.65(10^5)$ *Ans.* $\log N = 5.6675$

**8.** $N = 9.23(10^3)$

**9.** $q = \dfrac{3.41(10^4)}{7.34}$ *Ans.* $\log q = 3.6671$

**10.** $x = \dfrac{5.8(10^2)}{6.66}$

**11.** Verify Laws (II), (III), and (IV) above.

---

**8.** *The logarithm of any positive number $N$ (rounded off to 4 significant digits) can be determined with the aid of Table II by first expressing the number in the so-called standard form $N = n(10^k)$, where $n$ is between 1 and 10 and where $k$ is a positive or negative integer or is zero.*

### EXAMPLE 1

If $N = 634{,}000$, what is $\log N$?

**Solution:** In *standard form,*

$$N = 6.34(10^5)$$

Hence, by the method of Example (**b**) of the preceding article,

$$\begin{aligned}\log N &= \log 6.34 + \log (10^5)\\ &= 0.8021 + 5\\ &= 5.8021\end{aligned}$$

### EXAMPLE 2

Find log 0.05628.

**Solution:** In standard form,

$$0.05628 = 5.628(10^{-2})$$

Hence

$$\begin{aligned}\log 0.05628 &= \log 5.628 + \log (10^{-2})\\ &= 0.7504 - 2\end{aligned}$$

It proves convenient to write logarithms such as those found in Example 2 with the negative part kept separated from the positive part. In fact, many persons prefer to adjust such logarithms so that the negative part is $-10$. Thus, since $-2 = 8 - 10$,

$$\log 0.05628 = 8.7504 - 10$$

In composite logarithms such as found in the examples just given, the decimal part of the logarithm is called the *mantissa*, and the integral part the *characteristic*. Thus in

$$\log 634{,}000 = 5.8021$$

the mantissa of the logarithm is 0.8021 and the characteristic is 5. Again in

$$\log 0.05628 = 0.7504 - 2 \text{ (or } 8.7504 - 10)$$

the mantissa is 0.7504 and the characteristic is $-2$.

From the expression $N = n(10^k)$, the characteristic of $\log N$ is $k$ and the mantissa is $\log n$.

## *EXERCISES*

Making use of the answers given to the odd-numbered Exercises from 1 to 11, Paragraph 6, find the logarithms of the numbers below.

**1.** 365.5 — *Ans.* 2.5629
**2.** 36550
**3.** 0.003655 — *Ans.* 0.5629 − 3
**4.** 0.3655
**5.** 89.12 — *Ans.* 1.9500
**6.** $8.912(10^8)$
**7.** 0.08912 — *Ans.* 8.9500 − 10
**8.** 0.00008912
**9.** 2576 — *Ans.* 3.4109
**10.** 257600

With the aid of the tables find the logarithms of the following numbers, performing any intermediate calculations mentally.

**11.** 5930 — *Ans.* 3.7731
**12.** 47.4
**13.** 4.56 — *Ans.* 0.6590
**14.** 21
**15.** 0.005 — *Ans.* 7.6990 − 10
**16.** 8040
**17.** 101 — *Ans.* 2.0043
**18.** 0.387
**19.** 0.0712 — *Ans.* 0.8525 − 2
**20.** 7000

---

**9.** *If log N is given, N can be determined by reversing the steps suggested in Paragraph 8 for finding the logarithm of a number.* This will be illustrated with examples.

**EXAMPLE 1**

If $\log N = 3.8142$, what is $N$?

**Solution:** Since $\log N = 0.8142 + 3$, then

$$N = (\text{number whose log is } 0.8142)(\text{number whose log is } 3)$$
$$= 6.52(10^3) \quad [\text{from the tables}]$$
$$= 6520$$

**Alternative solution of 1:** Since $\log N = 3.8142$, then, by the definition of a logarithm,

$$\begin{aligned} N &= 10^{3.8142} \\ &= 10^{0.8142} \times 10^3 \\ &= 6.52(10^3) \quad \text{[from the tables]} \\ &= 6520 \end{aligned}$$

**EXAMPLE 2**

If $\log N = 7.7290 - 10$, what is $N$?

**Solution:** Since $\log N = 0.7290 - 3$, then

$$\begin{aligned} N &= (\text{number whose log is } 0.7290)\ (\text{number whose log is } -3) \\ &= 5.358(10^{-3}) \\ &= 0.005358 \end{aligned}$$

## *EXERCISES*

Using the answers to Exercises 13 to 23, Paragraph 6, find the value of $N$, given that

**1.** $\log N = 1.4806$ *Ans.* $N = 30.24$
**2.** $\log N = 9.4806 - 10$
**3.** $\log N = 3.4806$ *Ans.* $N = 3024$
**4.** $\log N = 2.4806$
**5.** $\log N = 0.6388 - 2$ *Ans.* $N = 0.04353$
**6.** $\log N = 1.6388$
**7.** $\log N = 3.6388$ *Ans.* $N = 4353$
**8.** $\log N = 7.6388 - 10$
**9.** $\log N = 9.8303 - 10$ *Ans.* $N = 0.6766$
**10.** $\log N = 2.8303$

With the aid of the tables, find $x$, given that

**11.** $\log x = 8.7574 - 10$ *Ans.* $x = 0.0572$
**12.** $\log x = 2.4082$
**13.** $\log x = 1.5132$ *Ans.* $x = 32.6$
**14.** $\log x = 9.2504 - 10$
**15.** $\log x = 0.9325$ *Ans.* $x = 8.56$
**16.** $\log x = 3.8222$
**17.** $\log x = 0.9046 - 3$ *Ans.* $x = 0.008028$
**18.** $\log x = 1.7828$
**19.** $\log x = 4.3724$ *Ans.* $x = 23570$
**20.** $\log x = 0.4545$

**10.** *Calculations involving multiplication, division, powers, and roots may be performed by means of simpler operations with the logarithms of the given numbers.* Paragraph 7 shows how to express the logarithm of a given expression involving these operations. Paragraphs 8 and 9 provide the information necessary to complete the calculation. Certain devices can be used to avoid obtaining *negative fractions* in the computations, as shown in Example 2 below. Negative fractions are avoided if the tables are to be used, since the entries in the tables are all positive.

## EXAMPLE 1

With logarithms find $\frac{(65.4)(1.657)^2}{(0.872)(43.6)}$

**Solution:** Let $N = \frac{(65.4)(1.657)^2}{(0.872)(43.6)}$. Then

$$\log N = \log 65.4 + 2 \log 1.657 - (\log 0.872 + \log 43.6)$$

A suggested arrangement of the details follows:

$$\begin{aligned} \log 1.657 &= 0.2193 \\ 2 \log 1.657 &= 0.4386 \\ \log 65.4 &= 1.8156 \quad (+) \\ \log (\text{numerator}) &= 2.2542 \\ \log (\text{denominator}) &= 1.5800 \quad (-) \\ \log N &= 0.6742 \end{aligned}$$

$$\begin{aligned} \log 0.872 &= 0.9405 - 1 \\ \log 43.6 &= 1.6395 \quad (+) \\ \log (\text{denom.}) &= 1.5800 \end{aligned}$$

and, from the tables,

$$N = 4.723$$

## EXAMPLE 2

Find $\sqrt[3]{\frac{2.304}{87.5}}$

**Solution:** Let $x = \sqrt[3]{\frac{2.304}{87.5}}$. Then

$$\log x = \tfrac{1}{3} (\log 2.304 - \log 87.5)$$

Computation:

$$\begin{aligned} \log 2.304 &= 0.3625 \\ \log 87.5 &= 1.9420 \quad (-) \end{aligned}$$

Subtraction in this form results in a negative fraction. To avoid this difficulty, simply add zero in the form 2 − 2 to the first logarithm, thus writing

0.3625 as 2.3625 − 2. Then

$$\begin{aligned} \log 2.304 &= 2.3625 - 2 \\ \log 87.5 &= 1.9420 \\ \log (\text{fraction}) &= \overline{0.4205 - 2} \quad (-) \end{aligned}$$

Now

$$\log x = \tfrac{1}{3}(0.4205 - 2)$$

Again, if the division is carried out in the present form, a negative decimal will result. This difficulty can be avoided by use of the equivalent form, with $0 = 1 - 1$ added in the parentheses,

$$\begin{aligned} \log x &= \tfrac{1}{3}(1.4205 - 3) \\ &= 0.4735 - 1 \end{aligned}$$

and, from the tables,

$$x = 0.2975$$

## *EXERCISES*

By means of logarithms calculate the values of the unknowns to four significant figures:

**1.** $N = 371.5 \times 4.27$ *Ans.* 1586

**2.** $x = 34.7 \times 484 \times 0.495$

**3.** $t = \dfrac{834}{6.43}$ *Ans.* 129.7

**4.** $n = \dfrac{5600}{0.034 \times 767}$

**5.** $x = \sqrt{84.69}$ *Ans.* 9.202

**6.** $r = \sqrt[3]{285}$

**7.** $y = (7.452)^3$ *Ans.* 413.9

**8.** $x = (17.5)^{3/2}$

**9.** $n = \dfrac{3.14}{88.4}$ *Ans.* 0.03551

**10.** $m = \dfrac{62.82}{0.372}$

**11.** $z = \sqrt{0.684}$ *Ans.* 0.8272

**12.** $n = \sqrt[3]{0.612}$

Solve each of the following equations, applying the following suggestions: (**a**) Take the logarithm of each side of the equation and simplify by the laws of logarithms, (**b**) solve the resulting equation for the logarithm of the unknown, and (**c**) evaluate the unknown as in the first twelve exercises.

**13.** (**a**) If $n$ is the number of shot 0.12 in. in radius that can be made from a cylinder of lead 6 in. long and 1.5 in. in radius, then

$$\frac{4\pi(0.12)^3 n}{3} = \pi(1.5)^2(6)$$

Find $n$.

(**b**) Find the number of shot 0.14 in. in *diameter* which can be made from a cylinder of lead 8 in. long and 2 in. in diameter.

**14.** $62.5x^3 = 546{,}000$ *Ans.* $x = 20.59$

**15.** $\dfrac{x}{0.5645} = \dfrac{67.5}{83.7}$

**16.** $x^2 = (4630)(781)$ *Ans.* $x = 1902$

**17.** $2.556 = 3.14\sqrt{\dfrac{l}{32.2}}$

**18.** $\frac{4}{3}(3.1416)r^3 = 205^3$ *Ans.* 127.2

**19.** $L = 0.492(.001189)(166)(95.4)^2$

# TABLES

## I. Squares, Square Roots, Cubes, and Cube Roots

| $n$ | $n^2$ | $n^3$ | $\sqrt{n}$ | $\sqrt[3]{n}$ | | $n^2$ | $n^3$ | $\sqrt{n}$ | $\sqrt[3]{n}$ |
|---|---|---|---|---|---|---|---|---|---|
| **1** | 1 | 1 | 1 | 1 | **51** | 2601 | 132651 | 7.141 | 3.708 |
| 2 | 4 | 8 | 1.414 | 1.260 | 52 | 2704 | 140608 | 7.211 | 3.733 |
| 3 | 9 | 27 | 1.732 | 1.442 | 53 | 2809 | 148877 | 7.280 | 3.756 |
| 4 | 16 | 64 | 2.000 | 1.587 | 54 | 2916 | 157464 | 7.348 | 3.780 |
| 5 | 25 | 125 | 2.236 | 1.710 | 55 | 3025 | 166375 | 7.416 | 3.803 |
| 6 | 36 | 216 | 2.449 | 1.817 | 56 | 3136 | 175616 | 7.483 | 3.826 |
| 7 | 49 | 343 | 2.646 | 1.913 | 57 | 3249 | 185193 | 7.550 | 3.849 |
| 8 | 64 | 512 | 2.828 | 2.000 | 58 | 3364 | 195112 | 7.616 | 3.871 |
| 9 | 81 | 729 | 3.000 | 2.080 | 59 | 3481 | 205379 | 7.681 | 3.893 |
| 10 | 100 | 1000 | 3.162 | 2.154 | 60 | 3600 | 216000 | 7.746 | 3.915 |
| **11** | 121 | 1331 | 3.317 | 2.224 | **61** | 3721 | 226981 | 7.810 | 3.936 |
| 12 | 144 | 1728 | 3.464 | 2.289 | 62 | 3844 | 238328 | 7.874 | 3.958 |
| 13 | 169 | 2197 | 3.606 | 2.351 | 63 | 3969 | 250047 | 7.937 | 3.979 |
| 14 | 196 | 2744 | 3.742 | 2.410 | 64 | 4096 | 262144 | 8.000 | 4.000 |
| 15 | 225 | 3375 | 3.873 | 2.466 | 65 | 4225 | 274625 | 8.062 | 4.021 |
| 16 | 256 | 4096 | 4.000 | 2.520 | 66 | 4356 | 287496 | 8.124 | 4.041 |
| 17 | 289 | 4913 | 4.123 | 2.571 | 67 | 4489 | 300763 | 8.185 | 4.062 |
| 18 | 324 | 5832 | 4.243 | 2.621 | 68 | 4624 | 314432 | 8.246 | 4.082 |
| 19 | 361 | 6859 | 4.359 | 2.668 | 69 | 4761 | 328509 | 8.307 | 4.102 |
| 20 | 400 | 8000 | 4.472 | 2.714 | 70 | 4900 | 343000 | 8.367 | 4.121 |
| **21** | 441 | 9261 | 4.583 | 2.759 | **71** | 5041 | 357911 | 8.426 | 4.141 |
| 22 | 484 | 10648 | 4.690 | 2.802 | 72 | 5184 | 373248 | 8.485 | 4.160 |
| 23 | 529 | 12167 | 4.796 | 2.844 | 73 | 5329 | 389017 | 8.544 | 4.179 |
| 24 | 576 | 13824 | 4.899 | 2.884 | 74 | 5476 | 405224 | 8.602 | 4.198 |
| 25 | 625 | 15625 | 5.000 | 2.924 | 75 | 5625 | 421875 | 8.660 | 4.217 |
| 26 | 676 | 17576 | 5.099 | 2.962 | 76 | 5776 | 438976 | 8.718 | 4.236 |
| 27 | 729 | 19683 | 5.196 | 3.000 | 77 | 5929 | 456533 | 8.775 | 4.254 |
| 28 | 784 | 21952 | 5.291 | 3.037 | 78 | 6084 | 474552 | 8.832 | 4.273 |
| 29 | 841 | 24389 | 5.385 | 3.072 | 79 | 6241 | 493039 | 8.888 | 4.291 |
| 30 | 900 | 27000 | 5.477 | 3.107 | 80 | 6400 | 512000 | 8.944 | 4.309 |
| **31** | 961 | 29791 | 5.568 | 3.141 | **81** | 6561 | 531441 | 9.000 | 4.327 |
| 32 | 1024 | 32768 | 5.657 | 3.175 | 82 | 6724 | 551368 | 9.055 | 4.344 |
| 33 | 1089 | 35937 | 5.745 | 3.208 | 83 | 6889 | 571787 | 9.110 | 4.362 |
| 34 | 1156 | 39304 | 5.831 | 3.240 | 84 | 7056 | 592704 | 9.165 | 4.380 |
| 35 | 1225 | 42875 | 5.916 | 3.271 | 85 | 7225 | 614125 | 9.220 | 4.397 |
| 36 | 1296 | 46656 | 6.000 | 3.302 | 86 | 7396 | 636056 | 9.274 | 4.414 |
| 37 | 1369 | 50653 | 6.083 | 3.332 | 87 | 7569 | 658503 | 9.327 | 4.431 |
| 38 | 1444 | 54872 | 6.164 | 3.362 | 88 | 7744 | 681472 | 9.381 | 4.448 |
| 39 | 1521 | 59319 | 6.245 | 3.391 | 89 | 7921 | 704969 | 9.434 | 4.465 |
| 40 | 1600 | 64000 | 6.325 | 3.420 | 90 | 8100 | 729000 | 9.487 | 4.481 |
| **41** | 1681 | 68921 | 6.403 | 3.448 | **91** | 8281 | 753571 | 9.539 | 4.498 |
| 42 | 1764 | 74088 | 6.481 | 3.476 | 92 | 8464 | 778688 | 9.592 | 4.514 |
| 43 | 1849 | 79507 | 6.557 | 3.503 | 93 | 8649 | 804357 | 9.644 | 4.531 |
| 44 | 1936 | 85184 | 6.633 | 3.530 | 94 | 8836 | 830584 | 9.695 | 4.547 |
| 45 | 2025 | 91125 | 6.708 | 3.557 | 95 | 9025 | 857375 | 9.747 | 4.563 |
| 46 | 2116 | 97336 | 6.782 | 3.583 | 96 | 9216 | 884736 | 9.798 | 4.579 |
| 47 | 2209 | 103823 | 6.856 | 3.609 | 97 | 9409 | 912673 | 9.849 | 4.595 |
| 48 | 2304 | 110592 | 6.928 | 3.634 | 98 | 9604 | 941192 | 9.899 | 4.610 |
| 49 | 2401 | 117649 | 7.000 | 3.659 | 99 | 9801 | 970299 | 9.950 | 4.626 |
| 50 | 2500 | 125000 | 7.071 | 3.684 | 100 | 10000 | 1000000 | 10.000 | 4.642 |
| $n$ | $n^2$ | $n^3$ | $\sqrt{n}$ | $\sqrt[3]{n}$ | $n$ | $n^2$ | $n^3$ | $\sqrt{n}$ | $\sqrt[3]{n}$ |

## II. Logarithms of Numbers

| No. | 0 | 1 | 2 | 3 | 4 | 5 | 6 | 7 | 8 | 9 |
|---|---|---|---|---|---|---|---|---|---|---|
| **10** | 0000 | 0043 | 0086 | 0128 | 0170 | 0212 | 0253 | 0294 | 0334 | 0374 |
| 11 | 0414 | 0453 | 0492 | 0531 | 0569 | 0607 | 0645 | 0682 | 0719 | 0755 |
| 12 | 0792 | 0828 | 0864 | 0899 | 0934 | 0969 | 1004 | 1038 | 1072 | 1106 |
| 13 | 1139 | 1173 | 1206 | 1239 | 1271 | 1303 | 1335 | 1367 | 1399 | 1430 |
| 14 | 1461 | 1492 | 1523 | 1553 | 1584 | 1614 | 1644 | 1673 | 1703 | 1732 |
| 15 | 1761 | 1790 | 1818 | 1847 | 1875 | 1903 | 1931 | 1959 | 1987 | 2014 |
| 16 | 2041 | 2068 | 2095 | 2122 | 2148 | 2175 | 2201 | 2227 | 2253 | 2279 |
| 17 | 2304 | 2330 | 2355 | 2380 | 2405 | 2430 | 2455 | 2480 | 2504 | 2529 |
| 18 | 2553 | 2577 | 2601 | 2625 | 2648 | 2672 | 2695 | 2718 | 2742 | 2765 |
| 19 | 2788 | 2810 | 2833 | 2856 | 2878 | 2900 | 2923 | 2945 | 2967 | 2989 |
| **20** | 3010 | 3032 | 3054 | 3075 | 3096 | 3118 | 3139 | 3160 | 3181 | 3201 |
| 21 | 3222 | 3243 | 3263 | 3284 | 3304 | 3324 | 3345 | 3365 | 3385 | 3404 |
| 22 | 3424 | 3444 | 3464 | 3483 | 3502 | 3522 | 3541 | 3560 | 3579 | 3598 |
| 23 | 3617 | 3636 | 3655 | 3674 | 3692 | 3711 | 3729 | 3747 | 3766 | 3784 |
| 24 | 3802 | 3820 | 3838 | 3856 | 3874 | 3892 | 3909 | 3927 | 3945 | 3962 |
| 25 | 3979 | 3997 | 4014 | 4031 | 4048 | 4065 | 4082 | 4099 | 4116 | 4133 |
| 26 | 4150 | 4166 | 4183 | 4200 | 4216 | 4232 | 4249 | 4265 | 4281 | 4298 |
| 27 | 4314 | 4330 | 4346 | 4362 | 4378 | 4393 | 4409 | 4425 | 4440 | 4456 |
| 28 | 4472 | 4487 | 4502 | 4518 | 4533 | 4548 | 4564 | 4579 | 4594 | 4609 |
| 29 | 4624 | 4639 | 4654 | 4669 | 4683 | 4698 | 4713 | 4728 | 4742 | 4757 |
| **30** | 4771 | 4786 | 4800 | 4814 | 4829 | 4843 | 4857 | 4871 | 4886 | 4900 |
| 31 | 4914 | 4928 | 4942 | 4955 | 4969 | 4983 | 4997 | 5011 | 5024 | 5038 |
| 32 | 5051 | 5065 | 5079 | 5092 | 5105 | 5119 | 5132 | 5145 | 5159 | 5172 |
| 33 | 5185 | 5198 | 5211 | 5224 | 5237 | 5250 | 5263 | 5276 | 5289 | 5302 |
| 34 | 5315 | 5328 | 5340 | 5353 | 5366 | 5378 | 5391 | 5403 | 5416 | 5428 |
| 35 | 5441 | 5453 | 5465 | 5478 | 5490 | 5502 | 5514 | 5527 | 5539 | 5551 |
| 36 | 5563 | 5575 | 5587 | 5599 | 5611 | 5623 | 5635 | 5647 | 5658 | 5670 |
| 37 | 5682 | 5694 | 5705 | 5717 | 5729 | 5740 | 5752 | 5763 | 5775 | 5786 |
| 38 | 5798 | 5809 | 5821 | 5832 | 5843 | 5855 | 5866 | 5877 | 5888 | 5899 |
| 39 | 5911 | 5922 | 5933 | 5944 | 5955 | 5966 | 5977 | 5988 | 5999 | 6010 |
| **40** | 6021 | 6031 | 6042 | 6053 | 6064 | 6075 | 6085 | 6096 | 6107 | 6117 |
| 41 | 6128 | 6138 | 6149 | 6160 | 6170 | 6180 | 6191 | 6201 | 6212 | 6222 |
| 42 | 6232 | 6243 | 6253 | 6263 | 6274 | 6284 | 6294 | 6304 | 6314 | 6325 |
| 43 | 6335 | 6345 | 6355 | 6365 | 6375 | 6385 | 6395 | 6405 | 6415 | 6425 |
| 44 | 6435 | 6444 | 6454 | 6464 | 6474 | 6484 | 6493 | 6503 | 6513 | 6522 |
| 45 | 6532 | 6542 | 6551 | 6561 | 6571 | 6580 | 6590 | 6599 | 6609 | 6618 |
| 46 | 6628 | 6637 | 6646 | 6656 | 6665 | 6675 | 6684 | 6693 | 6702 | 6712 |
| 47 | 6721 | 6730 | 6739 | 6749 | 6758 | 6767 | 6776 | 6785 | 6794 | 6803 |
| 48 | 6812 | 6821 | 6830 | 6839 | 6848 | 6857 | 6866 | 6875 | 6884 | 6893 |
| 49 | 6902 | 6911 | 6920 | 6928 | 6937 | 6946 | 6955 | 6964 | 6972 | 6981 |
| **50** | 6990 | 6998 | 7007 | 7016 | 7024 | 7033 | 7042 | 7050 | 7059 | 7067 |
| 51 | 7076 | 7084 | 7093 | 7101 | 7110 | 7118 | 7126 | 7135 | 7143 | 7152 |
| 52 | 7160 | 7168 | 7177 | 7185 | 7193 | 7202 | 7210 | 7218 | 7226 | 7235 |
| 53 | 7243 | 7251 | 7259 | 7267 | 7275 | 7284 | 7292 | 7300 | 7308 | 7316 |
| 54 | 7324 | 7332 | 7340 | 7348 | 7356 | 7364 | 7372 | 7380 | 7388 | 7396 |
| No. | 0 | 1 | 2 | 3 | 4 | 5 | 6 | 7 | 8 | 9 |

**Prop. Parts**

| | **43** | **42** |
|---|---|---|
| 1 | 4.3 | 4.2 |
| 2 | 8.6 | 8.4 |
| 3 | 12.9 | 12.6 |
| 4 | 17.2 | 16.8 |
| 5 | 21.5 | 21.0 |
| 6 | 25.8 | 25.2 |
| 7 | 30.1 | 29.4 |
| 8 | 34.4 | 33.6 |
| 9 | 38.7 | 37.8 |

| | **41** | **40** |
|---|---|---|
| 1 | 4.1 | 4.0 |
| 2 | 8.2 | 8.0 |
| 3 | 12.3 | 12.0 |
| 4 | 16.4 | 16.0 |
| 5 | 20.5 | 20.0 |
| 6 | 24.6 | 24.0 |
| 7 | 28.7 | 28.0 |
| 8 | 32.8 | 32.0 |
| 9 | 36.9 | 36.0 |

| | **39** | **38** |
|---|---|---|
| 1 | 3.9 | 3.8 |
| 2 | 7.8 | 7.6 |
| 3 | 11.7 | 11.4 |
| 4 | 15.6 | 15.2 |
| 5 | 19.5 | 19.0 |
| 6 | 23.4 | 22.8 |
| 7 | 27 3 | 26.6 |
| 8 | 31.2 | 30.4 |
| 9 | 35.1 | 34.2 |

| | **37** | **36** |
|---|---|---|
| 1 | 3.7 | 3.6 |
| 2 | 7.4 | 7.2 |
| 3 | 11.1 | 10.8 |
| 4 | 14.8 | 14.4 |
| 5 | 18.5 | 18.0 |
| 6 | 22.2 | 21.6 |
| 7 | 25.9 | 25.2 |
| 8 | 29.6 | 28.8 |
| 9 | 33.3 | 32.4 |

| | **35** | **34** |
|---|---|---|
| 1 | 3.5 | 3.4 |
| 2 | 7.0 | 6.8 |
| 3 | 10.5 | 10.2 |
| 4 | 14.0 | 13.6 |
| 5 | 17.5 | 17.0 |
| 6 | 21.0 | 20.4 |
| 7 | 24.5 | 23.8 |
| 8 | 28.0 | 27.2 |
| 9 | 31.5 | 30.6 |

| | **33** | **32** |
|---|---|---|
| 1 | 3.3 | 3.2 |
| 2 | 6.6 | 6.4 |
| 3 | 9.9 | 9.6 |
| 6 | 13.2 | 12.8 |
| 5 | 16.5 | 16.0 |
| 6 | 19.8 | 19.2 |
| 7 | 23.1 | 22.4 |
| 8 | 26.4 | 25.6 |
| 9 | 29.7 | 28.8 |

**Prop. Parts**

## II. Logarithms of Numbers

| No. | 0 | 1 | 2 | 3 | 4 | 5 | 6 | 7 | 8 | 9 |
|---|---|---|---|---|---|---|---|---|---|---|
| 55 | 7404 | 7412 | 7419 | 7427 | 7435 | 7443 | 7451 | 7459 | 7466 | 7474 |
| 56 | 7482 | 7490 | 7497 | 7505 | 7513 | 7520 | 7528 | 7536 | 7543 | 7551 |
| 57 | 7559 | 7566 | 7574 | 7582 | 7589 | 7597 | 7604 | 7612 | 7619 | 7627 |
| 58 | 7634 | 7642 | 7649 | 7657 | 7664 | 7672 | 7679 | 7686 | 7694 | 7701 |
| 59 | 7709 | 7716 | 7723 | 7731 | 7738 | 7745 | 7752 | 7760 | 7767 | 7774 |
| **60** | 7782 | 7789 | 7796 | 7803 | 7810 | 7818 | 7825 | 7832 | 7839 | 7846 |
| 61 | 7853 | 7860 | 7868 | 7875 | 7882 | 7889 | 7896 | 7903 | 7910 | 7917 |
| 62 | 7924 | 7931 | 7938 | 7945 | 7952 | 7959 | 7966 | 7973 | 7980 | 7987 |
| 63 | 7993 | 8000 | 8007 | 8014 | 8021 | 8028 | 8035 | 8041 | 8048 | 8055 |
| 64 | 8062 | 8069 | 8075 | 8082 | 8089 | 8096 | 8102 | 8109 | 8116 | 8122 |
| 65 | 8129 | 8136 | 8142 | 8149 | 8156 | 8162 | 8169 | 8176 | 8182 | 8189 |
| 66 | 8195 | 8202 | 8209 | 8215 | 8222 | 8228 | 8235 | 8241 | 8248 | 8254 |
| 67 | 8261 | 8267 | 8274 | 8280 | 8287 | 8293 | 8299 | 8306 | 8312 | 8319 |
| 68 | 8325 | 8331 | 8338 | 8344 | 8351 | 8357 | 8363 | 8370 | 8376 | 8382 |
| 69 | 8388 | 8395 | 8401 | 8407 | 8414 | 8420 | 8426 | 8432 | 8439 | 8445 |
| **70** | 8451 | 8457 | 8463 | 8470 | 8476 | 8482 | 8488 | 8494 | 8500 | 8506 |
| 71 | 8513 | 8519 | 8525 | 8531 | 8537 | 8543 | 8549 | 8555 | 8561 | 8567 |
| 72 | 8573 | 8579 | 8585 | 8591 | 8597 | 8603 | 8609 | 8615 | 8621 | 8627 |
| 73 | 8633 | 8639 | 8645 | 8651 | 8657 | 8663 | 8669 | 8675 | 8681 | 8686 |
| 74 | 8692 | 8698 | 8704 | 8710 | 8716 | 8722 | 8727 | 8733 | 8739 | 8745 |
| 75 | 8751 | 8756 | 8762 | 8768 | 8774 | 8779 | 8785 | 8791 | 8797 | 8802 |
| 76 | 8808 | 8814 | 8820 | 8825 | 8831 | 8837 | 8842 | 8848 | 8854 | 8859 |
| 77 | 8865 | 8871 | 8876 | 8882 | 8887 | 8893 | 8899 | 8904 | 8910 | 8915 |
| 78 | 8921 | 8927 | 8932 | 8938 | 8943 | 8949 | 8954 | 8960 | 8965 | 8971 |
| 79 | 8976 | 8982 | 8987 | 8993 | 8998 | 9004 | 9009 | 9015 | 9020 | 9025 |
| **80** | 9031 | 9036 | 9042 | 9047 | 9053 | 9058 | 9063 | 9069 | 9074 | 9079 |
| 81 | 9085 | 9090 | 9096 | 9101 | 9106 | 9112 | 9117 | 9122 | 9128 | 9133 |
| 82 | 9138 | 9143 | 9149 | 9154 | 9159 | 9165 | 9170 | 9175 | 9180 | 9186 |
| 83 | 9191 | 9196 | 9201 | 9206 | 9212 | 9217 | 9222 | 9227 | 9232 | 9238 |
| 84 | 9243 | 9248 | 9253 | 9258 | 9263 | 9269 | 9274 | 9279 | 9284 | 9289 |
| 85 | 9294 | 9299 | 9304 | 9309 | 9315 | 9320 | 9325 | 9330 | 9335 | 9340 |
| 86 | 9345 | 9350 | 9355 | 9360 | 9365 | 9370 | 9375 | 9380 | 9385 | 9390 |
| 87 | 9395 | 9400 | 9405 | 9410 | 9415 | 9420 | 9425 | 9430 | 9435 | 9440 |
| 88 | 9445 | 9450 | 9455 | 9460 | 9465 | 9469 | 9474 | 9479 | 9484 | 9489 |
| 89 | 9494 | 9499 | 9504 | 9509 | 9513 | 9518 | 9523 | 9528 | 9533 | 9538 |
| **90** | 9542 | 9547 | 9552 | 9557 | 9562 | 9566 | 9571 | 9576 | 9581 | 9586 |
| 91 | 9590 | 9595 | 9600 | 9605 | 9609 | 9614 | 9619 | 9624 | 9628 | 9633 |
| 92 | 9638 | 9643 | 9647 | 9652 | 9657 | 9661 | 9666 | 9671 | 9675 | 9680 |
| 93 | 9685 | 9689 | 9694 | 9699 | 9703 | 9708 | 9713 | 9717 | 9722 | 9727 |
| 94 | 9731 | 9736 | 9741 | 9745 | 9750 | 9754 | 9759 | 9763 | 9768 | 9773 |
| 95 | 9777 | 9782 | 9786 | 9791 | 9795 | 9800 | 9805 | 9809 | 9814 | 9818 |
| 96 | 9823 | 9827 | 9832 | 9836 | 9841 | 9845 | 9850 | 9854 | 9859 | 9863 |
| 97 | 9868 | 9872 | 9877 | 9881 | 9886 | 9890 | 9894 | 9899 | 9903 | 9908 |
| 98 | 9912 | 9917 | 9921 | 9926 | 9930 | 9934 | 9939 | 9943 | 9948 | 9952 |
| 99 | 9956 | 9961 | 9965 | 9969 | 9974 | 9978 | 9983 | 9987 | 9991 | 9996 |
| No. | 0 | 1 | 2 | 3 | 4 | 5 | 6 | 7 | 8 | 9 |

**Prop. Parts**

| | 31 | 30 | 29 | 28 | 27 | 26 | 25 | 24 | 23 | 22 | 21 |
|---|---|---|---|---|---|---|---|---|---|---|---|
| 1 | 3.1 | 3.0 | 2.9 | 2.8 | 2.7 | 2.6 | 2.5 | 2.4 | 2.3 | 2.2 | 2.1 |
| 2 | 6.2 | 6.0 | 5.8 | 5.6 | 5.4 | 5.2 | 5.0 | 4.8 | 4.6 | 4.4 | 4.2 |
| 3 | 9.3 | 9.0 | 8.7 | 8.4 | 8.1 | 7.8 | 7.5 | 7.2 | 6.9 | 6.6 | 6.3 |
| 4 | 12.4 | 12.0 | 11.6 | 11.2 | 10.8 | 10.4 | 10.0 | 9.6 | 9.2 | 8.8 | 8.4 |
| 5 | 15.5 | 15.0 | 14.5 | 14.0 | 13.5 | 13.0 | 12.5 | 12.0 | 11.5 | 11.0 | 10.5 |
| 6 | 18.6 | 18.0 | 17.4 | 16.8 | 16.2 | 15.6 | 15.0 | 14.4 | 13.8 | 13.2 | 12.6 |
| 7 | 21.7 | 21.0 | 20.3 | 19.6 | 18.9 | 18.2 | 17.5 | 16.8 | 16.1 | 15.4 | 14.7 |
| 8 | 24.8 | 24.0 | 23.2 | 22.4 | 21.6 | 20.8 | 20.0 | 19.2 | 18.4 | 17.6 | 16.8 |
| 9 | 27.9 | 27.0 | 26.1 | 25.2 | 24.3 | 23.4 | 22.5 | 21.6 | 20.7 | 19.8 | 18.9 |

Prop. Parts

## III. Natural Functions

| $x$ | sin $x$ | cos $x$ | tan $x$ | cot $x$ | sec $x$ | cosec $x$ | |
|---|---|---|---|---|---|---|---|
| **0° 0′** | .00000 | 1.0000 | .00000 | ∞ | 1.0000 | ∞ | **90° 0′** |
| 10′ | .00291 | 1.0000 | .00291 | 343.77 | 1.0000 | 343.78 | 50′ |
| 20′ | .00582 | 1.0000 | .00582 | 171.88 | 1.0000 | 171.89 | 40′ |
| 30′ | .00873 | 1.0000 | .00873 | 114.59 | 1.0000 | 114.59 | 30′ |
| 40′ | .01164 | .9999 | .01164 | 85.940 | 1.0001 | 85.946 | 20′ |
| 50′ | .01454 | .9999 | .01455 | 68.750 | 1.0001 | 68.757 | 10′ |
| **1° 0′** | .01745 | .9998 | .01746 | 57.290 | 1.0002 | 57.299 | **89° 0′** |
| 10′ | .02036 | .9998 | .02036 | 49.104 | 1.0002 | 49.114 | 50′ |
| 20′ | .02327 | .9997 | .02328 | 42.964 | 1.0003 | 42.976 | 40′ |
| 30′ | .02618 | .9997 | .02619 | 38.188 | 1.0003 | 38.202 | 30′ |
| 40′ | .02908 | .9996 | .02910 | 34.368 | 1.0004 | 34.382 | 20′ |
| 50′ | .03199 | .9995 | .03201 | 31.242 | 1.0005 | 31.258 | 10′ |
| **2° 0′** | .03490 | .9994 | .03492 | 28.6363 | 1.0006 | 28.654 | **88° 0′** |
| 10′ | .03781 | .9993 | .03783 | 26.4316 | 1.0007 | 26.451 | 50′ |
| 20′ | .04071 | .9992 | .04075 | 24.5418 | 1.0008 | 24.562 | 40′ |
| 30′ | .04362 | .9990 | .04366 | 22.9038 | 1.0010 | 22.926 | 30′ |
| 40′ | .04653 | .9989 | .04658 | 21.4704 | 1.0011 | 21.494 | 20′ |
| 50′ | .04943 | .9988 | .04949 | 20.2056 | 1.0012 | 20.230 | 10′ |
| **3° 0′** | .05234 | .9986 | .05241 | 19.0811 | 1.0014 | 19.107 | **87° 0′** |
| 10′ | .05524 | .9985 | .05533 | 18.0750 | 1.0015 | 18.103 | 50′ |
| 20′ | .05814 | .9983 | .05824 | 17.1693 | 1.0017 | 17.198 | 40′ |
| 30′ | .06105 | .9981 | .06116 | 16.3499 | 1.0019 | 16.380 | 30′ |
| 40′ | .06395 | .9980 | .06408 | 15.6048 | 1.0021 | 15.637 | 20′ |
| 50′ | .06685 | .9978 | .06700 | 14.9244 | 1.0022 | 14.958 | 10′ |
| **4° 0′** | .06976 | .9976 | .06993 | 14.3007 | 1.0024 | 14.336 | **86° 0′** |
| 10′ | .07266 | .9974 | .07285 | 13.7267 | 1.0027 | 13.763 | 50′ |
| 20′ | .07556 | .9971 | .07578 | 13.1969 | 1.0029 | 13.235 | 40′ |
| 30′ | .07846 | .9969 | .07870 | 12.7062 | 1.0031 | 12.746 | 30′ |
| 40′ | .08136 | .9967 | .08163 | 12.2505 | 1.0033 | 12.291 | 20′ |
| 50′ | .08426 | .9964 | .08456 | 11.8262 | 1.0036 | 11.868 | 10′ |
| **5° 0′** | .08716 | .9962 | .08749 | 11.4301 | 1.0038 | 11.474 | **85° 0′** |
| 10′ | .09005 | .9959 | .09042 | 11.0594 | 1.0041 | 11.105 | 50′ |
| 20′ | .09295 | .9957 | .09335 | 10.7119 | 1.0044 | 10.758 | 40′ |
| 30′ | .09585 | .9954 | .09629 | 10.3854 | 1.0046 | 10.433 | 30′ |
| 40′ | .09874 | .9951 | .09923 | 10.0780 | 1.0049 | 10.128 | 20′ |
| 50′ | .10164 | .9948 | .10216 | 9.7882 | 1.0052 | 9.839 | 10′ |
| **6° 0′** | .10453 | .9945 | .10510 | 9.5144 | 1.0055 | 9.5668 | **84° 0′** |
| 10′ | .10742 | .9942 | .10805 | 9.2553 | 1.0058 | 9.3092 | 50′ |
| 20′ | .11031 | .9939 | .11099 | 9.0098 | 1.0061 | 9.0652 | 40′ |
| 30′ | .11320 | .9936 | .11394 | 8.7769 | 1.0065 | 8.8337 | 30′ |
| 40′ | .11609 | .9932 | .11688 | 8.5555 | 1.0068 | 8.6138 | 20′ |
| 50′ | .11898 | .9929 | .11983 | 8.3450 | 1.0072 | 8.4647 | 10′ |
| **7° 0′** | .12187 | .9925 | .12278 | 8.1443 | 1.0075 | 8.2055 | **83° 0′** |
| 10′ | .12476 | .9922 | .12574 | 7.9530 | 1.0079 | 8.0157 | 50′ |
| 20′ | .12764 | .9918 | .12869 | 7.7704 | 1.0083 | 7.8344 | 40′ |
| 30′ | .13053 | .9914 | .13165 | 7.5958 | 1.0086 | 7.6613 | 30′ |
| | cos $x$ | sin $x$ | cot $x$ | tan $x$ | cosec $x$ | sec $x$ | $x$ |

# III. Natural Functions

| *x* | | sin *x* | cos *x* | tan *x* | cot *x* | sec *x* | cosec *x* | | |
|---|---|---|---|---|---|---|---|---|---|
| | 30′ | .1305 | .9914 | .1317 | 7.5958 | 1.0086 | 7.6613 | | 30′ |
| | 40′ | .1334 | .9911 | .1346 | 7.4287 | 1.0090 | 7.4957 | | 20′ |
| | 50′ | .1363 | .9907 | .1376 | 7.2687 | 1.0094 | 7.3372 | | 10′ |
| **8°** | **0′** | .1392 | .9903 | .1405 | 7.1154 | 1.0098 | 7.1853 | **82°** | **0′** |
| | 10′ | .1421 | .9899 | .1435 | 6.9682 | 1.0102 | 7.0396 | | 50′ |
| | 20′ | .1449 | .9894 | .1465 | 6.8269 | 1.0107 | 6.8998 | | 40′ |
| | 30′ | .1478 | .9890 | .1495 | 6.6912 | 1.0111 | 6.7655 | | 30′ |
| | 40′ | .1507 | .9886 | .1524 | 6.5606 | 1.0116 | 6.6363 | | 20′ |
| | 50′ | .1536 | .9881 | .1554 | 6.4348 | 1.0120 | 6.5121 | | 10′ |
| **9°** | **0′** | .1564 | .9877 | .1584 | 6.3138 | 1.0125 | 6.3925 | **81°** | **0′** |
| | 10′ | .1593 | .9872 | .1614 | 6.1970 | 1.0129 | 6.2772 | | 50′ |
| | 20′ | .1622 | .9868 | .1644 | 6.0844 | 1.0134 | 6.1661 | | 40′ |
| | 30′ | .1650 | .9863 | .1673 | 5.9758 | 1.0139 | 6.0589 | | 30′ |
| | 40′ | .1679 | .9858 | .1703 | 5.8708 | 1.0144 | 5.9554 | | 20′ |
| | 50′ | .1708 | .9853 | .1733 | 5.7694 | 1.0149 | 5.8554 | | 10′ |
| **10°** | **0′** | .1736 | .9848 | .1763 | 5.6713 | 1.0154 | 5.7588 | **80°** | **0′** |
| | 10′ | .1765 | .9843 | .1793 | 5.5764 | 1.0160 | 5.6653 | | 50′ |
| | 20′ | .1794 | .9838 | .1823 | 5.4845 | 1.0165 | 5.5749 | | 40′ |
| | 30′ | .1822 | .9833 | .1853 | 5.3955 | 1.0170 | 5.4874 | | 30′ |
| | 40′ | .1851 | .9827 | .1883 | 5.3093 | 1.0176 | 5.4026 | | 20′ |
| | 50′ | .1880 | .9822 | .1914 | 5.2257 | 1.0182 | 5.3205 | | 10′ |
| **11°** | **0′** | .1908 | .9816 | .1944 | 5.1446 | 1.0187 | 5.2408 | **79°** | **0′** |
| | 10′ | .1937 | .9811 | .1974 | 5.0658 | 1.0193 | 5.1636 | | 50′ |
| | 20′ | .1965 | .9805 | .2004 | 4.9894 | 1.0199 | 5.0886 | | 40′ |
| | 30′ | .1994 | .9799 | .2035 | 4.9152 | 1.0205 | 5.0159 | | 30′ |
| | 40′ | .2022 | .9793 | .2065 | 4.8430 | 1.0211 | 4.9452 | | 20′ |
| | 50′ | .2051 | .9787 | .2095 | 4.7729 | 1.0217 | 4.8765 | | 10′ |
| **12°** | **0′** | .2079 | .9781 | .2126 | 4.7046 | 1.0223 | 4.8097 | **78°** | **0′** |
| | 10′ | .2108 | .9775 | .2156 | 4.6382 | 1.0230 | 4.7448 | | 50′ |
| | 20′ | .2136 | .9769 | .2186 | 4.5736 | 1.0236 | 4.6817 | | 40′ |
| | 30′ | .2164 | .9763 | .2217 | 4.5107 | 1.0243 | 4.6202 | | 30′ |
| | 40′ | .2193 | .9757 | .2247 | 4.4494 | 1.0249 | 4.5604 | | 20′ |
| | 50′ | .2221 | .9750 | .2278 | 4.3897 | 1.0256 | 4.5022 | | 10′ |
| **13°** | **0′** | .2250 | .9744 | .2309 | 4.3315 | 1.0263 | 4.4454 | **77°** | **0′** |
| | 10′ | .2278 | .9737 | .2339 | 4.2747 | 1.0270 | 4.3901 | | 50′ |
| | 20′ | .2306 | .9730 | .2370 | 4.2193 | 1.0277 | 4.3362 | | 40′ |
| | 30′ | .2334 | .9724 | .2401 | 4.1653 | 1.0284 | 4.2837 | | 30′ |
| | 40′ | .2363 | .9717 | .2432 | 4.1126 | 1.0291 | 4.2324 | | 20′ |
| | 50′ | .2391 | .9710 | .2462 | 4.0611 | 1.0299 | 4.1824 | | 10′ |
| **14°** | **0′** | .2419 | .9703 | .2493 | 4.0108 | 1.0306 | 4.1336 | **76°** | **0′** |
| | 10′ | .2447 | .9696 | .2524 | 3.9617 | 1.0314 | 4.0859 | | 50′ |
| | 20′ | .2476 | .9689 | .2555 | 3.9136 | 1.0321 | 4.0394 | | 40′ |
| | 30′ | .2504 | .9681 | .2586 | 3.8667 | 1.0329 | 3.9939 | | 30′ |
| | 40′ | .2532 | .9674 | .2617 | 3.8208 | 1.0337 | 3.9495 | | 20′ |
| | 50′ | .2560 | .9667 | .2648 | 3.7760 | 1.0345 | 3.9061 | | 10′ |
| **15°** | **0′** | .2588 | .9659 | .2679 | 3.7321 | 1.0353 | 3.8637 | **75°** | **0′** |
| | | cos *x* | sin *x* | cot *x* | tan *x* | cosec *x* | sec *x* | *x* | |

## III. Natural Functions

| $x$ | sin $x$ | cos $x$ | tan $x$ | cot $x$ | sec $x$ | cosec $x$ | |
|---|---|---|---|---|---|---|---|
| **15° 0′** | .2588 | .9659 | .2679 | 3.7321 | 1.0353 | 3.8637 | **75° 0′** |
| 10′ | .2616 | .9652 | .2711 | 3.6891 | 1.0361 | 3.8222 | 50′ |
| 20′ | .2644 | .9644 | .2742 | 3.6470 | 1.0369 | 3.7817 | 40′ |
| 30′ | .2672 | .9636 | .2773 | 3.6059 | 1.0377 | 3.7420 | 30′ |
| 40′ | .2700 | .9628 | .2805 | 3.5656 | 1.0386 | 3.7032 | 20′ |
| 50′ | .2728 | .9621 | .2836 | 3.5261 | 1.0394 | 3.6652 | 10′ |
| **16° 0′** | .2756 | .9613 | .2867 | 3.4874 | 1.0403 | 3.6280 | **74° 0′** |
| 10′ | .2784 | .9605 | .2899 | 3.4495 | 1.0412 | 3.5915 | 50′ |
| 20′ | .2812 | .9596 | .2931 | 3.4124 | 1.0421 | 3.5559 | 40′ |
| 30′ | .2840 | .9588 | .2962 | 3.3759 | 1.0430 | 3.5209 | 30′ |
| 40′ | .2868 | .9580 | .2994 | 3.3402 | 1.0439 | 3.4867 | 20′ |
| 50′ | .2896 | .9572 | .3026 | 3.3052 | 1.0448 | 3.4532 | 10′ |
| **17° 0′** | .2924 | .9563 | .3057 | 3.2709 | 1.0457 | 3.4203 | **73° 0′** |
| 10′ | .2952 | .9555 | .3089 | 3.2371 | 1.0466 | 3.3881 | 50′ |
| 20′ | .2979 | .9546 | .3121 | 3.2041 | 1.0476 | 3.3565 | 40′ |
| 30′ | .3007 | .9537 | .3153 | 3.1716 | 1.0485 | 3.3255 | 30′ |
| 40′ | .3035 | .9528 | .3185 | 3.1397 | 1.0495 | 3.2951 | 20′ |
| 50′ | .3062 | .9520 | .3217 | 3.1084 | 1.0505 | 3.2653 | 10′ |
| **18° 0′** | .3090 | .9511 | .3249 | 3.0777 | 1.0515 | 3.2361 | **72° 0′** |
| 10′ | .3118 | .9502 | .3281 | 3.0475 | 1.0525 | 3.2074 | 50′ |
| 20′ | .3145 | .9492 | 3314 | 3.0178 | 1.0535 | 3.1792 | 40′ |
| 30′ | .3173 | .9483 | .3346 | 2.9887 | 1.0545 | 3.1516 | 30′ |
| 40′ | .3201 | .9474 | .3378 | 2.9600 | 1.0555 | 3.1244 | 20′ |
| 50′ | .3228 | .9465 | .3411 | 2.9319 | 1.0566 | 3.0977 | 10′ |
| **19° 0′** | .3256 | .9455 | .3443 | 2.9042 | 1.0576 | 3.0716 | **71° 0′** |
| 10′ | .3283 | .9446 | .3476 | 2.8770 | 1.0587 | 3.0458 | 50′ |
| 20′ | .3311 | .9436 | .3508 | 2.8502 | 1.0598 | 3.0206 | 40′ |
| 30′ | .3338 | .9426 | .3541 | 2.8239 | 1.0609 | 2.9957 | 30′ |
| 40′ | .3365 | .9417 | .3574 | 2.7980 | 1.0620 | 2.9714 | 20′ |
| 50′ | .3393 | .9407 | .3607 | 2.7725 | 1.0631 | 2.9474 | 10′ |
| **20° 0′** | .3420 | .9397 | .3640 | 2.7475 | 1.0642 | 2.9238 | **70° 0′** |
| 10′ | .3448 | .9387 | .3673 | 2.7228 | 1.0653 | 2.9006 | 50′ |
| 20′ | .3475 | .9377 | .3706 | 2.6985 | 1.0665 | 2.8779 | 40′ |
| 30′ | .3502 | .9367 | .3739 | 2.6746 | 1.0676 | 2.8555 | 30′ |
| 40′ | .3529 | .9356 | .3772 | 2.6511 | 1.0688 | 2.8334 | 20′ |
| 50′ | .3557 | .9346 | .3805 | 2.6279 | 1.0700 | 2.8118 | 10′ |
| **21° 0′** | .3584 | .9336 | .3839 | 2.6051 | 1.0712 | 2.7904 | **69° 0′** |
| 10′ | .3611 | .9325 | .3872 | 2.5826 | 1.0724 | 2.7695 | 50′ |
| 20′ | .3638 | .9315 | .3906 | 2.5605 | 1.0736 | 2.7488 | 40′ |
| 30′ | .3665 | .9304 | .3939 | 2.5386 | 1.0748 | 2.7285 | 30′ |
| 40′ | .3692 | .9293 | .3973 | 2.5172 | 1.0760 | 2.7085 | 20′ |
| 50′ | .3719 | .9283 | .4006 | 2.4960 | 1.0773 | 2.6888 | 10′ |
| **22° 0′** | .3746 | .9272 | .4040 | 2.4751 | 1.0785 | 2.6695 | **68° 0′** |
| 10′ | .3773 | .9261 | .4074 | 2.4545 | 1.0798 | 2.6504 | 50′ |
| 20′ | .3800 | .9250 | .4108 | 2.4342 | 1.0811 | 2.6316 | 40′ |
| 30′ | .3827 | .9239 | .4142 | 2.4142 | 1.0824 | 2 6131 | 30′ |
| | cos $x$ | sin $x$ | cot $x$ | tan $x$ | cosec $x$ | sec $x$ | $x$ |

## III. Natural Functions

| *x* | sin *x* | cos *x* | tan *x* | cot *x* | sec *x* | cosec *x* | |
|---|---|---|---|---|---|---|---|
| 30′ | .3827 | .9239 | .4142 | 2.4142 | 1.0824 | 2.6131 | 30′ |
| 40′ | .3854 | .9228 | .4176 | 2.3945 | 1.0837 | 2.5949 | 20′ |
| 50′ | .3881 | .9216 | .4210 | 2.3750 | 1.0850 | 2.5770 | 10′ |
| **23° 0′** | .3907 | .9205 | .4245 | 2.3559 | 1.0864 | 2.5593 | **67° 0′** |
| 10′ | .3934 | .9194 | .4279 | 2.3369 | 1.0877 | 2.5419 | 50′ |
| 20′ | .3961 | .9182 | .4314 | 2.3183 | 1.0891 | 2.5247 | 40′ |
| 30′ | .3987 | .9171 | .4348 | 2.2998 | 1.0904 | 2.5078 | 30′ |
| 40′ | .4014 | .9159 | .4383 | 2.2817 | 1.0918 | 2.4912 | 20′ |
| 50′ | .4041 | .9147 | .4417 | 2.2637 | 1.0932 | 2.4748 | 10′ |
| **24° 0′** | .4067 | .9135 | .4452 | 2.2460 | 1.0946 | 2.4586 | **66° 0′** |
| 10′ | .4094 | .9124 | .4487 | 2.2286 | 1.0961 | 2.4426 | 50′ |
| 20′ | .4120 | .9112 | .4522 | 2.2113 | 1.0975 | 2.4269 | 40′ |
| 30′ | .4147 | .9100 | .4557 | 2.1943 | 1.0990 | 2.4114 | 30′ |
| 40′ | .4173 | .9088 | .4592 | 2.1775 | 1.1004 | 2.3961 | 20′ |
| 50′ | .4200 | .9075 | .4628 | 2.1609 | 1.1019 | 2.3811 | 10′ |
| **25° 0′** | .4226 | .9063 | .4663 | 2.1445 | 1.1034 | 2.3662 | **65° 0′** |
| 10′ | .4253 | .9051 | .4699 | 2.1283 | 1.1049 | 2.3515 | 50′ |
| 20′ | .4279 | .9038 | .4734 | 2.1123 | 1.1064 | 2.3371 | 40′ |
| 30′ | .4305 | .9026 | .4770 | 2.0965 | 1.1079 | 2.3228 | 30′ |
| 40′ | .4331 | .9013 | .4806 | 2.0809 | 1.1095 | 2.3088 | 20′ |
| 50′ | .4358 | .9001 | .4841 | 2.0655 | 1.1110 | 2.2949 | 10′ |
| **26° 0′** | .4384 | .8988 | .4877 | 2.0503 | 1.1126 | 2.2812 | **64° 0′** |
| 10′ | .4410 | .8975 | .4913 | 2.0353 | 1.1142 | 2.2677 | 50′ |
| 20′ | .4436 | .8962 | .4950 | 2.0204 | 1.1158 | 2.2543 | 40′ |
| 30′ | .4462 | .8949 | .4986 | 2.0057 | 1.1174 | 2.2412 | 30′ |
| 40′ | .4488 | .8936 | .5022 | 1.9912 | 1.1190 | 2.2282 | 20′ |
| 50′ | .4514 | .8923 | .5059 | 1.9768 | 1.1207 | 2.2154 | 10′ |
| **27° 0′** | .4540 | .8910 | .5095 | 1.9626 | 1.1223 | 2.2027 | **63° 0′** |
| 10′ | .4566 | .8897 | .5132 | 1.9486 | 1.1240 | 2.1902 | 50′ |
| 20′ | .4592 | .8884 | .5169 | 1.9347 | 1.1257 | 2.1779 | 40′ |
| 30′ | .4617 | .8870 | .5206 | 1.9210 | 1.1274 | 2.1657 | 30′ |
| 40′ | .4643 | .8857 | .5243 | 1.9074 | 1.1291 | 2.1537 | 20′ |
| 50′ | .4669 | .8843 | .5280 | 1.8940 | 1.1308 | 2.1418 | 10′ |
| **28° 0′** | .4695 | .8829 | .5317 | 1.8807 | 1.1326 | 2.1301 | **62° 0′** |
| 10′ | .4720 | .8816 | .5354 | 1.8676 | 1.1343 | 2.1185 | 50′ |
| 20′ | .4746 | .8802 | .5392 | 1.8546 | 1.1361 | 2.1070 | 40′ |
| 30′ | .4772 | .8788 | .5430 | 1.8418 | 1.1379 | 2.0957 | 30′ |
| 40′ | .4797 | .8774 | .5467 | 1.8291 | 1.1397 | 2.0846 | 20′ |
| 50′ | .4823 | .8760 | .5505 | 1.8165 | 1.1415 | 2.0736 | 10′ |
| **29° 0′** | .4848 | .8746 | .5543 | 1.8040 | 1.1434 | 2.0627 | **61° 0′** |
| 10′ | .4874 | .8732 | .5581 | 1.7917 | 1.1452 | 2.0519 | 50′ |
| 20′ | .4899 | .8718 | .5619 | 1.7796 | 1.1471 | 2.0413 | 40′ |
| 30′ | .4924 | .8704 | .5658 | 1.7675 | 1.1490 | 2.0308 | 30′ |
| 40′ | .4950 | .8689 | .5696 | 1.7556 | 1.1509 | 2.0204 | 20′ |
| 50′ | .4975 | .8675 | .5735 | 1.7437 | 1.1528 | 2.0101 | 10′ |
| **30° 0′** | .5000 | .8660 | .5774 | 1.7321 | 1.1547 | 2.0000 | **60° 0′** |
| | cos *x* | sin *x* | cot *x* | tan *x* | cosec *x* | sec *x* | *x* |

## III. Natural Functions

| $x$ | | sin $x$ | cos $x$ | tan $x$ | cot $x$ | sec $x$ | cosec $x$ | | |
|---|---|---|---|---|---|---|---|---|---|
| **30°** | **0′** | .5000 | .8660 | .5774 | 1.7321 | 1.1547 | 2.0000 | **60°** | **0′** |
| | 10′ | .5025 | .8646 | .5812 | 1.7205 | 1.1567 | 1.9900 | | 50′ |
| | 20′ | .5050 | .8631 | .5851 | 1.7090 | 1.1586 | 1.9801 | | 40′ |
| | 30′ | .5075 | .8616 | .5890 | 1.6977 | 1.1606 | 1.9703 | | 30′ |
| | 40′ | .5100 | .8601 | .5930 | 1.6864 | 1.1626 | 1.9606 | | 20′ |
| | 50′ | .5125 | .8587 | .5969 | 1.6753 | 1.1646 | 1.9511 | | 10′ |
| **31°** | **0′** | .5150 | .8572 | .6009 | 1.6643 | 1.1666 | 1.9416 | **59°** | **0′** |
| | 10′ | .5175 | .8557 | .6048 | 1.6534 | 1.1687 | 1.9323 | | 50′ |
| | 20′ | .5200 | .8542 | .6088 | 1.6426 | 1.1708 | 1.9230 | | 40′ |
| | 30′ | .5225 | .8526 | .6128 | 1.6319 | 1.1728 | 1.9139 | | 30′ |
| | 40′ | .5250 | .8511 | .6168 | 1.6212 | 1.1749 | 1.9049 | | 20′ |
| | 50′ | .5275 | .8496 | .6208 | 1.6107 | 1.1770 | 1.8959 | | 10′ |
| **32°** | **0′** | .5299 | .8480 | .6249 | 1.6003 | 1.1792 | 1.8871 | **58°** | **0′** |
| | 10′ | .5324 | .8465 | .6289 | 1.5900 | 1.1813 | 1.8783 | | 50′ |
| | 20′ | .5348 | .8450 | .6330 | 1.5798 | 1.1835 | 1.8699 | | 40′ |
| | 30′ | .5373 | .8434 | .6371 | 1.5697 | 1.1857 | 1.8612 | | 30′ |
| | 40′ | .5398 | .8418 | .6412 | 1.5597 | 1.1879 | 1.8527 | | 20′ |
| | 50′ | .5422 | .8403 | .6453 | 1.5497 | 1.1901 | 1.8444 | | 10′ |
| **33°** | **0′** | .5446 | .8387 | .6494 | 1.5399 | 1.1924 | 1.8361 | **57°** | **0′** |
| | 10′ | .5471 | .8371 | .6536 | 1.5301 | 1.1946 | 1.8279 | | 50′ |
| | 20′ | .5495 | .8355 | .6577 | 1.5204 | 1.1969 | 1.8198 | | 40′ |
| | 30′ | .5519 | .8339 | .6619 | 1.5108 | 1.1992 | 1.8118 | | 30′ |
| | 40′ | .5544 | .8323 | .6661 | 1.5013 | 1.2015 | 1.8039 | | 20′ |
| | 50′ | .5568 | .8307 | .6703 | 1.4919 | 1.2039 | 1 7960 | | 10′ |
| **34°** | **0′** | .5592 | .8290 | .6745 | 1.4826 | 1.2062 | 1.7883 | **56°** | **0′** |
| | 10′ | .5616 | .8274 | .6787 | 1.4733 | 1.2086 | 1.7806 | | 50′ |
| | 20′ | .5640 | .8258 | .6830 | 1.4641 | 1.2110 | 1.7730 | | 40′ |
| | 30′ | .5664 | .8241 | .6873 | 1.4550 | 1.2134 | 1.7655 | | 30′ |
| | 40′ | .5688 | .8225 | .6916 | 1.4460 | 1.2158 | 1.7581 | | 20′ |
| | 50′ | .5712 | .8208 | .6959 | 1.4370 | 1.2183 | 1.7507 | | 10′ |
| **35°** | **0′** | .5736 | .8192 | .7002 | 1.4281 | 1.2208 | 1.7435 | **55°** | **0′** |
| | 10′ | .5760 | .8175 | .7046 | 1.4193 | 1.2233 | 1.7362 | | 50′ |
| | 20′ | .5783 | .8158 | .7089 | 1.4106 | 1.2258 | 1.7291 | | 40′ |
| | 30′ | .5807 | .8141 | .7133 | 1.4019 | 1.2283 | 1.7221 | | 30′ |
| | 40′ | .5831 | .8124 | .7177 | 1.3934 | 1.2309 | 1.7151 | | 20′ |
| | 50′ | .5854 | .8107 | .7221 | 1.3848 | 1.2335 | 1.7082 | | 10′ |
| **36°** | **0′** | .5878 | .8090 | .7265 | 1.3764 | 1.2361 | 1.7013 | **54°** | **0′** |
| | 10′ | .5901 | .8073 | .7310 | 1.3680 | 1.2387 | 1.6945 | | 50′ |
| | 20′ | .5925 | .8056 | .7355 | 1.3597 | 1.2413 | 1.6878 | | 40′ |
| | 30′ | .5948 | .8039 | .7400 | 1.3514 | 1.2440 | 1.6812 | | 30′ |
| | 40′ | .5972 | .8021 | .7445 | 1.3432 | 1.2467 | 1.6746 | | 20′ |
| | 50′ | .5995 | .8004 | .7490 | 1.3351 | 1.2494 | 1.6681 | | 10′ |
| **37°** | **0′** | .6018 | .7986 | .7536 | 1.3270 | 1.2521 | 1.6616 | **53°** | **0′** |
| | 10′ | .6041 | .7969 | .7581 | 1.3190 | 1.2549 | 1.6553 | | 50′ |
| | 20′ | .6065 | .7951 | .7627 | 1.3111 | 1.2577 | 1.6489 | | 40′ |
| | 30′ | .6088 | .7934 | .7673 | 1.3032 | 1.2605 | 1.6427 | | 30′ |
| | | cos $x$ | sin $x$ | cot $x$ | tan $x$ | cosec $x$ | sec $x$ | $x$ | |

## III. Natural Functions

| $x$ | $\sin x$ | $\cos x$ | $\tan x$ | $\cot x$ | $\sec x$ | $\operatorname{cosec} x$ | |
|---|---|---|---|---|---|---|---|
| 30′ | .6088 | .7934 | .7673 | 1.3032 | 1.2605 | 1.6427 | 30′ |
| 40′ | .6111 | .7916 | .7720 | 1.2954 | 1.2633 | 1.6365 | 20′ |
| 50′ | .6134 | .7898 | .7766 | 1.2876 | 1.2662 | 1.6304 | 10′ |
| **38° 0′** | .6157 | .7880 | .7813 | 1.2799 | 1.2690 | 1.6243 | **52° 0′** |
| 10′ | .6180 | .7862 | .7860 | 1.2723 | 1.2719 | 1.6183 | 50′ |
| 20′ | .6202 | .7844 | .7907 | 1.2647 | 1.2748 | 1.6123 | 40′ |
| 30′ | .6225 | .7826 | .7954 | 1.2572 | 1.2779 | 1.6064 | 30′ |
| 40′ | .6248 | .7808 | .8002 | 1.2497 | 1.2808 | 1.6005 | 20′ |
| 50′ | .6271 | .7790 | .8050 | 1.2423 | 1.2837 | 1.5948 | 10′ |
| **39° 0′** | .6293 | .7771 | .8098 | 1.2349 | 1.2868 | 1.5890 | **51° 0′** |
| 10′ | .6316 | .7753 | .8146 | 1.2276 | 1.2898 | 1.5833 | 50′ |
| 20′ | .6338 | .7735 | .8195 | 1.2203 | 1.2929 | 1.5777 | 40′ |
| 30′ | .6361 | .7716 | .8243 | 1.2131 | 1.2960 | 1.5721 | 30′ |
| 40′ | .6383 | .7698 | .8292 | 1.2059 | 1.2991 | 1.5666 | 20′ |
| 50′ | .6406 | .7679 | .8342 | 1.1988 | 1.3022 | 1.5611 | 10′ |
| **40° 0′** | .6428 | .7660 | .8391 | 1.1918 | 1.3054 | 1.5557 | **50° 0′** |
| 10′ | .6450 | .7642 | .8441 | 1.1847 | 1.3086 | 1.5504 | 50′ |
| 20′ | .6472 | .7623 | .8491 | 1.1778 | 1.3118 | 1.5450 | 40′ |
| 30′ | .6494 | .7604 | .8541 | 1.1708 | 1 3151 | 1.5398 | 30′ |
| 40′ | .6517 | .7585 | .8591 | 1.1640 | 1.3184 | 1.5346 | 20′ |
| 50′ | .6539 | .7566 | .8642 | 1.1571 | 1.3217 | 1.5294 | 10′ |
| **41° 0′** | .6561 | .7547 | .8693 | 1.1504 | 1.3250 | 1.5243 | **49° 0′** |
| 10′ | .6583 | .7528 | .8744 | 1.1436 | 1.3284 | 1.5192 | 50′ |
| 20′ | .6604 | .7509 | .8796 | 1.1369 | 1.3318 | 1.5142 | 40′ |
| 30′ | .6626 | .7490 | .8847 | 1.1303 | 1.3352 | 1.5092 | 30′ |
| 40′ | .6648 | .7470 | .8899 | 1.1237 | 1.3386 | 1.5042 | 20′ |
| 50′ | .6670 | .7451 | .8952 | 1.1171 | 1.3421 | 1.4993 | 10′ |
| **42° 0′** | .6691 | .7431 | .9004 | 1.1106 | 1.3456 | 1.4945 | **48° 0′** |
| 10′ | .6713 | .7412 | .9057 | 1.1041 | 1.3492 | 1.4897 | 50′ |
| 20′ | .6734 | .7392 | .9110 | 1.0977 | 1.3527 | 1.4849 | 40′ |
| 30′ | .6756 | .7373 | .9163 | 1.0913 | 1.3563 | 1.4802 | 30′ |
| 40′ | .6777 | .7353 | .9217 | 1.0850 | 1.3600 | 1.4755 | 20′ |
| 50′ | .6799 | .7333 | .9271 | 1.0786 | 1.3636 | 1.4709 | 10′ |
| **43° 0′** | .6820 | .7314 | .9325 | 1.0724 | 1.3673 | 1.4663 | **47° 0′** |
| 10′ | .6841 | .7294 | .9380 | 1.0661 | 1.3711 | 1.4617 | 50′ |
| 20′ | .6862 | .7274 | .9435 | 1.0599 | 1.3748 | 1.4572 | 40′ |
| 30′ | .6884 | .7254 | .9490 | 1.0538 | 1.3786 | 1.4527 | 30′ |
| 40′ | .6905 | .7234 | .9545 | 1.0477 | 1.3824 | 1.4483 | 20′ |
| 50′ | .6926 | .7214 | .9601 | 1.0416 | 1.3863 | 1.4439 | 10′ |
| **44° 0′** | .6947 | .7193 | .9657 | 1.0355 | 1.3902 | 1.4396 | **46° 0′** |
| 10′ | .6967 | .7173 | .9713 | 1.0295 | 1.3941 | 1.4352 | 50′ |
| 20′ | .6988 | .7153 | .9770 | 1.0235 | 1.3980 | 1.4310 | 40′ |
| 30′ | .7009 | .7133 | .9827 | 1.0176 | 1.4020 | 1.4267 | 30′ |
| 40′ | .7030 | .7112 | .9884 | 1.0117 | 1.4061 | 1.4225 | 20′ |
| 50′ | .7050 | .7092 | .9942 | 1.0058 | 1.4101 | 1.4184 | 10′ |
| **45° 0′** | .7071 | .7071 | 1.0000 | 1.0000 | 1.4142 | 1.4142 | **45° 0′** |
| | $\cos x$ | $\sin x$ | $\cot x$ | $\tan x$ | $\operatorname{cosec} x$ | $\sec x$ | $x$ |

## IV. Logarithms of Sines, Cosines,

| x | log sin | d | log cos | d | log tan | d | log cot | |
|---|---|---|---|---|---|---|---|---|
| 0° 0′ | −∞ | | 10.0000 | | −∞ | | ∞ | 90° 0′ |
| 10′ | 7.4637 | | .0000 | 0 | 7.4637 | | 2.5363 | 50′ |
| 20′ | .7648 | 3011 | .0000 | 0 | .7648 | 3011 | .2352 | 40′ |
| 30′ | .9408 | 1760 | .0000 | 0 | .9409 | 1761 | .0591 | 30′ |
| 40′ | 8.0658 | 1250 | .0000 | 0 | 8.0658 | 1249 | 1.9342 | 20′ |
| 50′ | .1627 | 969 | .0000 | 0 | .1627 | 969 | .8373 | 10′ |
| 1° 0′ | 8.2419 | 792 | 9.9999 | 1 | 8.2419 | 792 | 1.7581 | 89° 0′ |
| 10′ | .3088 | 669 | .9999 | 0 | .3089 | 670 | .6911 | 50′ |
| 20′ | .3668 | 580 | .9999 | 0 | .3669 | 580 | .6331 | 40′ |
| 30′ | .4179 | 511 | 9.9999 | 0 | .4181 | 512 | .5819 | 30′ |
| 40′ | .4637 | 458 | .9998 | 1 | .4638 | 457 | .5362 | 20′ |
| 50′ | .5050 | 413 | .9998 | 0 | .5053 | 415 | .4947 | 10′ |
| 2° 0′ | 8.5428 | 378 | 9.9997 | 1 | 8.5431 | 378 | 1.4569 | 88° 0′ |
| 10′ | .5776 | 348 | .9997 | 0 | .5779 | 348 | .4221 | 50′ |
| 20′ | .6097 | 321 | .9996 | 1 | .6101 | 322 | .3899 | 40′ |
| 30′ | .6397 | 300 | .9996 | 0 | .6401 | 300 | .3599 | 30′ |
| 40′ | .6677 | 280 | .9995 | 1 | .6682 | 281 | .3318 | 20′ |
| 50′ | .6940 | 263 | .9995 | 0 | .6945 | 263 | .3055 | 10′ |
| 3° 0′ | 8.7188 | 248 | 9.9994 | 1 | 8.7194 | 249 | 1.2806 | 87° 0′ |
| 10′ | .7423 | 235 | .9993 | 1 | .7429 | 235 | .2571 | 50′ |
| 20′ | .7645 | 222 | .9993 | 0 | .7652 | 223 | .2348 | 40′ |
| 30′ | .7857 | 212 | .9992 | 1 | .7865 | 213 | .2135 | 30′ |
| 40′ | .8059 | 202 | .9991 | 1 | .8067 | 202 | .1933 | 20′ |
| 50′ | .8251 | 192 | .9990 | 1 | .8261 | 194 | .1739 | 10′ |
| 4° 0′ | 8.8436 | 185 | 9.9989 | 1 | 8.8446 | 185 | 1.1554 | 86° 0′ |
| 10′ | .8613 | 177 | .9989 | 0 | .8624 | 178 | .1376 | 50′ |
| 20′ | .8783 | 170 | .9988 | 1 | .8795 | 171 | .1205 | 40′ |
| 30′ | .8946 | 163 | .9987 | 1 | .8960 | 165 | .1040 | 30′ |
| 40′ | .9104 | 158 | .9986 | 1 | .9118 | 158 | .0882 | 20′ |
| 50′ | .9256 | 152 | .9985 | 1 | .9272 | 154 | .0728 | 10′ |
| 5° 0′ | 8.9403 | 147 | 9.9983 | 2 | 8.9420 | 148 | 1.0580 | 85° 0′ |
| 10′ | .9545 | 142 | .9982 | 1 | .9563 | 143 | .0437 | 50′ |
| 20′ | .9682 | 137 | .9981 | 1 | .9701 | 138 | .0299 | 40′ |
| 30′ | .9816 | 134 | .9980 | 1 | .9836 | 135 | .0164 | 30′ |
| 40′ | .9945 | 129 | .9979 | 1 | .9966 | 130 | .0034 | 20′ |
| 50′ | 9.0070 | 125 | .9977 | 2 | 9.0093 | 127 | 0.9907 | 10′ |
| 6° 0′ | 9.0192 | 122 | 9.9976 | 1 | 9.0216 | 123 | 0.9784 | 84° 0′ |
| 10′ | .0311 | 119 | .9975 | 1 | .0336 | 120 | .9664 | 50′ |
| 20′ | .0426 | 115 | .9973 | 2 | .0453 | 117 | .9547 | 40′ |
| 30′ | .0539 | 113 | .9972 | 1 | .0567 | 114 | .9433 | 30′ |
| 40′ | .0648 | 109 | .9971 | 1 | .0678 | 111 | .9322 | 20′ |
| 50′ | .0755 | 107 | .9969 | 2 | .0786 | 108 | .9214 | 10′ |
| 7° 0′ | 9.0859 | 104 | 9.9968 | 1 | 9.0891 | 105 | 0.9109 | 83° 0′ |
| 10′ | .0961 | 102 | .9966 | 2 | .0995 | 104 | .9005 | 50′ |
| 20′ | .1060 | 99 | .9964 | 2 | .1096 | 101 | .8904 | 40′ |
| 30′ | .1157 | 97 | .9963 | 1 | .1194 | 98 | .8806 | 30′ |
| | log cos | d | log sin | d | log cot | d | log tan | x |

**Small Angles**

| x | S | T |
|---|---|---|
| <1° | 6.4637 | 6.4637 |
| 1° | 6.4637 | 6.4638 |
| 2° | 6.4636 | 6.4639 |
| 3° | 6.4635 | 6.4641 |
| 4° | 6.4634 | 6.4644 |
| 5° | 6.4631 | 6.4649 |

**Prop. Parts.**

| | 113 | 111 | 109 |
|---|---|---|---|
| 1 | 11.3 | 11.1 | 10.9 |
| 2 | 22.6 | 22.2 | 21.8 |
| 3 | 33.9 | 33.3 | 32.7 |
| 4 | 45.2 | 44.4 | 43.6 |
| 5 | 56.5 | 55.5 | 54.5 |
| 6 | 67.8 | 66.6 | 65.4 |
| 7 | 79.1 | 77.7 | 76.3 |
| 8 | 90.4 | 88.8 | 87.2 |
| 9 | 101.7 | 99.9 | 98.1 |

| | 108 | 107 | 105 |
|---|---|---|---|
| 1 | 10.8 | 10.7 | 10.5 |
| 2 | 21.6 | 21.4 | 21.0 |
| 3 | 32.4 | 32.1 | 31.5 |
| 4 | 43.2 | 42.8 | 42.0 |
| 5 | 54.0 | 53.5 | 52.5 |
| 6 | 64.8 | 64.2 | 63.0 |
| 7 | 75.6 | 74.9 | 73.5 |
| 8 | 86.4 | 85.6 | 84.0 |
| 9 | 97.2 | 96.3 | 94.5 |

| | 104 | 102 | 101 |
|---|---|---|---|
| 1 | 10.4 | 10.2 | 10.1 |
| 2 | 20.8 | 20.4 | 20.2 |
| 3 | 31.2 | 30.6 | 30.3 |
| 4 | 41.6 | 40.8 | 40.4 |
| 5 | 52.0 | 51.0 | 50.5 |
| 6 | 62.4 | 61.2 | 60.6 |
| 7 | 72.8 | 71.4 | 70.7 |
| 8 | 83.2 | 81.6 | 80.8 |
| 9 | 93.6 | 91.8 | 90.9 |

| | 99 | 98 | 97 | 95 |
|---|---|---|---|---|
| 1 | 9.9 | 9.8 | 9.7 | 9.5 |
| 2 | 19.8 | 19.6 | 19.4 | 19.0 |
| 3 | 29.7 | 29.4 | 29.1 | 28.5 |
| 4 | 39.6 | 39.2 | 38.8 | 38.0 |
| 5 | 49.5 | 49.0 | 48.5 | 47.5 |
| 6 | 59.4 | 58.8 | 58.2 | 57.0 |
| 7 | 69.3 | 68.6 | 67.9 | 66.5 |
| 8 | 79.2 | 78.4 | 77.6 | 76.0 |
| 9 | 89.1 | 88.2 | 87.3 | 85.5 |

| | 143 | 142 | 138 | 137 | 135 | 134 | 130 | 129 | 127 | 125 | 123 | 122 | 119 | 117 | 115 | 114 |
|---|---|---|---|---|---|---|---|---|---|---|---|---|---|---|---|---|
| 1 | 14.3 | 14.2 | 13.8 | 13.7 | 13.5 | 13.4 | 13.0 | 12.9 | 12.7 | 12.5 | 12.3 | 12.2 | 11.9 | 11.7 | 11.5 | 11.4 |
| 2 | 28.6 | 28.4 | 27.6 | 27.4 | 27.0 | 26.8 | 26.0 | 25.8 | 25.4 | 25.0 | 24.6 | 24.4 | 23.8 | 23.4 | 23.0 | 22.8 |
| 3 | 42.9 | 42.6 | 41.4 | 41.1 | 40.5 | 40.2 | 39.0 | 38.7 | 38.1 | 37.5 | 36.9 | 36.6 | 35.7 | 35.1 | 34.5 | 34.2 |
| 4 | 57.2 | 56.8 | 55.2 | 54.8 | 54.0 | 53.6 | 52.0 | 51.6 | 50.8 | 50.0 | 49.2 | 48.8 | 47.6 | 46.8 | 46.0 | 45.6 |
| 5 | 71.5 | 71.0 | 69.0 | 68.5 | 67.5 | 67.0 | 65.0 | 64.5 | 63.5 | 62.5 | 61.5 | 61.0 | 59.5 | 58.5 | 57.5 | 57.0 |
| 6 | 85.8 | 85.2 | 82.8 | 82.2 | 81.0 | 80.4 | 78.0 | 77.4 | 76.2 | 75.0 | 73.8 | 73.2 | 71.4 | 70.2 | 69.0 | 68.4 |
| 7 | 100.1 | 99.4 | 96.6 | 95.9 | 94.5 | 93.8 | 91.0 | 90.3 | 88.9 | 87.5 | 86.1 | 85.4 | 83.3 | 81.9 | 80.5 | 79.8 |
| 8 | 114.4 | 113.6 | 110.4 | 109.6 | 108.0 | 107.2 | 104.0 | 103.2 | 101.6 | 100.0 | 98.4 | 97.6 | 95.2 | 93.6 | 92.0 | 91.2 |
| 9 | 128.7 | 127.8 | 124.2 | 123.3 | 121.5 | 120.6 | 117.0 | 116.1 | 114.3 | 112.5 | 110.7 | 109.8 | 107.1 | 105.3 | 103.5 | 102.6 |

## Tangents, and Cotangents. IV

| x | log sin | d | log cos | d | log tan | d | log cot | |
|---|---|---|---|---|---|---|---|---|
| 30′ | 9.1157 | 95 | 9.9963 | 2 | 9.1194 | 97 | 0.8806 | 30′ |
| 40′ | .1252 | 93 | .9961 | 2 | .1291 | 94 | .8709 | 20′ |
| 50′ | .1345 | 91 | .9959 | 1 | .1385 | 93 | .8615 | 10′ |
| **8° 0′** | 9.1436 | 89 | 9.9958 | 2 | 9.1478 | 91 | 0.8522 | **82° 0′** |
| 10′ | .1525 | 87 | .9956 | 2 | .1569 | 89 | .8431 | 50′ |
| 20′ | .1612 | 85 | .9954 | 2 | .1658 | 87 | .8342 | 40′ |
| 30′ | .1697 | 84 | .9952 | 2 | .1745 | 86 | .8255 | 30′ |
| 40′ | .1781 | 82 | .9950 | 2 | .1831 | 84 | .8169 | 20′ |
| 50′ | .1863 | 80 | .9948 | 2 | .1915 | 82 | .8085 | 10′ |
| **9° 0′** | 9.1943 | 79 | 9.9946 | 2 | 9.1997 | 81 | 0.8003 | **81° 0′** |
| 10′ | .2022 | 78 | .9944 | 2 | .2078 | 80 | .7922 | 50′ |
| 20′ | .2100 | 76 | .9942 | 2 | .2158 | 78 | .7842 | 40′ |
| 30′ | .2176 | 75 | .9940 | 2 | .2236 | 77 | .7764 | 30′ |
| 40′ | .2251 | 73 | .9938 | 2 | .2313 | 76 | .7687 | 20′ |
| 50′ | .2324 | 73 | .9936 | 2 | .2389 | 74 | .7611 | 10′ |
| **10° 0′** | 9.2397 | 71 | 9.9934 | 3 | 9.2463 | 73 | 0.7537 | **80° 0′** |
| 10′ | .2468 | 70 | .9931 | 2 | .2536 | 73 | .7464 | 50′ |
| 20′ | .2538 | 68 | .9929 | 2 | .2609 | 71 | .7391 | 40′ |
| 30′ | .2606 | 68 | .9927 | 3 | .2680 | 70 | .7320 | 30′ |
| 40′ | .2674 | 66 | .9924 | 2 | .2750 | 69 | .7250 | 20′ |
| 50′ | .2740 | 66 | .9922 | 3 | .2819 | 68 | .7181 | 10′ |
| **11° 0′** | 9.2806 | 64 | 9.9919 | 2 | 9.2887 | 66 | 0.7113 | **79° 0′** |
| 10′ | .2870 | 64 | .9917 | 3 | .2953 | 67 | .7047 | 50′ |
| 20′ | .2934 | 63 | .9914 | 2 | .3020 | 65 | .6980 | 40′ |
| 30′ | .2997 | 61 | .9912 | 3 | .3085 | 64 | .6915 | 30′ |
| 40′ | .3058 | 61 | .9909 | 2 | .3149 | 63 | .6851 | 20′ |
| 50′ | .3119 | 60 | .9907 | 3 | .3212 | 63 | .6788 | 10′ |
| **12° 0′** | 9.3179 | 59 | 9.9904 | 3 | 9.3275 | 61 | 0.6725 | **78° 0′** |
| 10′ | .3238 | 58 | .9901 | 2 | .3336 | 61 | .6664 | 50′ |
| 20′ | .3296 | 57 | .9899 | 3 | .3397 | 61 | .6603 | 40′ |
| 30′ | .3353 | 57 | .9896 | 3 | .3458 | 59 | .6542 | 30′ |
| 40′ | .3410 | 56 | .9893 | 3 | .3517 | 59 | .6483 | 20′ |
| 50′ | .3466 | 55 | .9890 | 3 | .3576 | 58 | .6424 | 10′ |
| **13° 0′** | 9.3521 | 54 | 9.9887 | 3 | 9.3634 | 57 | 0.6366 | **77° 0′** |
| 10′ | .3575 | 54 | .9884 | 3 | .3691 | 57 | .6309 | 50′ |
| 20′ | .3629 | 53 | .9881 | 3 | .3748 | 56 | .6252 | 40′ |
| 30′ | .3682 | 52 | .9878 | 3 | .3804 | 55 | .6196 | 30′ |
| 40′ | .3734 | 52 | .9875 | 3 | .3859 | 55 | .6141 | 20′ |
| 50′ | .3786 | 51 | .9872 | 3 | .3914 | 54 | .6086 | 10′ |
| **14° 0′** | 9.3837 | 50 | 9.9869 | 3 | 9.3968 | 53 | 0.6032 | **76° 0′** |
| 10′ | .3887 | 50 | .9866 | 3 | .4021 | 53 | .5979 | 50′ |
| 20′ | .3937 | 49 | .9863 | 4 | .4074 | 53 | .5926 | 40′ |
| 30′ | .3986 | 49 | .9859 | 3 | .4127 | 51 | .5873 | 30′ |
| 40′ | .4035 | 48 | .9856 | 3 | .4178 | 52 | .5822 | 20′ |
| 50′ | .4083 | 47 | .9853 | 4 | .4230 | 51 | .5770 | 10′ |
| **15° 0′** | 9.4130 | | 9.9849 | | 9.4281 | | 0.5719 | **75° 0′** |
| | log cos | d | log sin | d | log cot | d | log tan | x |

**Prop. Parts**

| | 73 | 71 | 70 | 69 | 68 |
|---|---|---|---|---|---|
| 1 | 7.3 | 7.1 | 7.0 | 6.9 | 6.8 |
| 2 | 14.6 | 14.2 | 14.0 | 13.8 | 13.6 |
| 3 | 21.9 | 21.3 | 21.0 | 20.7 | 20.4 |
| 4 | 29.2 | 28.4 | 28.0 | 27.6 | 27.2 |
| 5 | 36.5 | 35.5 | 35.0 | 34.5 | 34.0 |
| 6 | 43.8 | 42.6 | 42.0 | 41.4 | 40.8 |
| 7 | 51.1 | 49.7 | 49.0 | 48.3 | 47.6 |
| 8 | 58.4 | 56.8 | 56.0 | 55.2 | 54.4 |
| 9 | 65.7 | 63.9 | 63.0 | 62.1 | 61.2 |

| | 67 | 66 | 65 | 64 | 63 |
|---|---|---|---|---|---|
| 1 | 6.7 | 6.6 | 6.5 | 6.4 | 6 3 |
| 2 | 13.4 | 13.2 | 13.0 | 12.8 | 12.6 |
| 3 | 20.1 | 19.8 | 19.5 | 19.2 | 18.9 |
| 4 | 26.8 | 26,4 | 26.0 | 25.6 | 25.2 |
| 5 | 33.5 | 33.0 | 32.5 | 32.0 | 31.5 |
| 6 | 40.2 | 39.6 | 39.0 | 38.4 | 37.8 |
| 7 | 46.9 | 46.2 | 45.5 | 44.8 | 44.1 |
| 8 | 53.6 | 52.8 | 52.0 | 51.2 | 50.4 |
| 9 | 60.3 | 59.4 | 58.5 | 57.6 | 56.7 |

| | 61 | 60 | 59 | 58 | 57 |
|---|---|---|---|---|---|
| 1 | 6.1 | 6.0 | 5.9 | 5.8 | 5.7 |
| 2 | 12.2 | 12.0 | 11.8 | 11.6 | 11.4 |
| 3 | 18.3 | 18.0 | 17.7 | 17.4 | 17.1 |
| 4 | 24.4 | 24.0 | 23.6 | 23.2 | 22.8 |
| 5 | 30.5 | 30.0 | 29.5 | 29.0 | 28.5 |
| 6 | 36.6 | 36.0 | 35.4 | 34.8 | 34.2 |
| 7 | 42.7 | 42.0 | 41.3 | 40.6 | 39.9 |
| 8 | 48.8 | 48.0 | 47.2 | 46.4 | 45.6 |
| 9 | 54.9 | 54.0 | 53.1 | 52.2 | 51.3 |

| | 56 | 55 | 54 | 53 | 52 |
|---|---|---|---|---|---|
| 1 | 5.6 | 5.5 | 5.4 | 5.3 | 5.2 |
| 2 | 11.2 | 11.0 | 10.8 | 10.6 | 10.4 |
| 3 | 16.8 | 16.5 | 16.2 | 15.9 | 15.6 |
| 4 | 22.4 | 22.0 | 21.6 | 21.2 | 20.8 |
| 5 | 28.0 | 27.5 | 27.0 | 26.5 | 26.0 |
| 6 | 33.6 | 33.0 | 32.4 | 31.8 | 31.2 |
| 7 | 39.2 | 38.5 | 37.8 | 37.1 | 36.4 |
| 8 | 44.8 | 44.0 | 43.2 | 42.4 | 41.6 |
| 9 | 50.4 | 49.5 | 48.6 | 47.7 | 46.8 |

| | 51 | 50 | 48 | 47 |
|---|---|---|---|---|
| 1 | 5.1 | 5.0 | 4.8 | 4.7 |
| 2 | 10.2 | 10.0 | 9.6 | 9.4 |
| 3 | 15.3 | 15.0 | 14.4 | 14.1 |
| 4 | 20.4 | 20.0 | 19.2 | 18.8 |
| 5 | 25.5 | 25.0 | 24.0 | 23.5 |
| 6 | 30.6 | 30.0 | 28.8 | 28.2 |
| 7 | 35.7 | 35.0 | 33.6 | 32.9 |
| 8 | 40.8 | 40.0 | 38.4 | 37.6 |
| 9 | 45.9 | 45.5 | 43.2 | 42.3 |

| | 97 | 94 | 93 | 91 | 89 | 87 | 86 | 85 | 84 | 82 | 81 | 79 | 78 | 77 | 76 | 75 | 74 |
|---|---|---|---|---|---|---|---|---|---|---|---|---|---|---|---|---|---|
| 1 | 9.7 | 9.4 | 9.3 | 9.1 | 8.9 | 8.7 | 8.6 | 8.5 | 8.4 | 8.2 | 8.1 | 7.9 | 7.8 | 7.7 | 7.6 | 7.5 | 7.4 |
| 2 | 19.4 | 18.8 | 18.6 | 18.2 | 17.8 | 17.4 | 17.2 | 17.0 | 16.8 | 16.4 | 16.2 | 15.8 | 15.6 | 15.4 | 15.2 | 15.0 | 14.8 |
| 3 | 29.1 | 28.2 | 27.9 | 27.3 | 26.7 | 26.1 | 25.8 | 25.5 | 25.2 | 24.6 | 24.3 | 23.7 | 23.4 | 23.1 | 22.8 | 22.5 | 22.2 |
| 4 | 38.8 | 37.6 | 37.2 | 36.4 | 35.6 | 34.8 | 34.4 | 34.0 | 33.6 | 32.8 | 32.4 | 31.6 | 31.2 | 30.8 | 30.4 | 30.0 | 29.6 |
| 5 | 48.5 | 47.0 | 46.5 | 45.5 | 44.5 | 43.5 | 43.0 | 42.5 | 42.0 | 41.0 | 40.5 | 39.5 | 39.0 | 38.5 | 38.0 | 37.5 | 37.0 |
| 6 | 58.2 | 56.4 | 55.8 | 54.6 | 53.4 | 52.2 | 51.6 | 51.0 | 50.4 | 49.2 | 48.6 | 47.4 | 46.8 | 46.2 | 45.6 | 45.0 | 44.4 |
| 7 | 67.9 | 65.8 | 65.1 | 63.7 | 62.3 | 60.9 | 60.2 | 59.5 | 58.8 | 57.4 | 56.7 | 55.3 | 54.6 | 53.9 | 53.2 | 52.5 | 51.8 |
| 8 | 77.6 | 75.2 | 74.4 | 72.8 | 71.2 | 69.6 | 68.8 | 68.0 | 67.2 | 65.6 | 64.8 | 63.2 | 62.4 | 61.6 | 60.8 | 60.0 | 59.2 |
| 9 | 87.3 | 84.6 | 83.7 | 81.9 | 80.1 | 78.3 | 77.4 | 76.5 | 75.6 | 73.8 | 72.9 | 71.1 | 70.2 | 69.3 | 68.4 | 67.5 | 66.6 |

## IV. Logarithms of Sines, Cosines,

| $x$ | log sin | d | log cos | d | log tan | d | log cot | |
|---|---|---|---|---|---|---|---|---|
| 15° 0′ | 9.4130 | 47 | 9.9849 | 3 | 9.4281 | 50 | 0.5719 | 75° 0′ |
| 10′ | .4177 | 46 | .9846 | 3 | .4331 | 50 | .5669 | 50′ |
| 20′ | .4223 | 46 | .9843 | 4 | .4381 | 49 | .5619 | 40′ |
| 30′ | .4269 | 45 | .9839 | 3 | .4430 | 49 | .5570 | 30′ |
| 40′ | .4314 | 45 | .9836 | 4 | .4479 | 48 | .5521 | 20′ |
| 50′ | .4359 | 44 | .9832 | 4 | .4527 | 48 | .5473 | 10′ |
| 16° 0′ | 9.4403 | 44 | 9.9828 | 3 | 9.4575 | 47 | 0.5425 | 74° 0′ |
| 10′ | .4447 | 44 | .9825 | 4 | .4622 | 47 | .5378 | 50′ |
| 20′ | .4491 | 42 | .9821 | 4 | .4669 | 47 | .5331 | 40′ |
| 30′ | .4533 | 43 | .9817 | 3 | .4716 | 46 | .5284 | 30′ |
| 40′ | .4576 | 42 | .9814 | 4 | .4762 | 46 | .5238 | 20′ |
| 50′ | .4618 | 41 | .9810 | 4 | .4808 | 45 | .5192 | 10′ |
| 17° 0′ | 9.4659 | 41 | 9.9806 | 4 | 9.4853 | 45 | 0.5147 | 73° 0′ |
| 10′ | .4700 | 41 | .9802 | 4 | .4898 | 45 | .5102 | 50′ |
| 20′ | .4741 | 40 | .9798 | 4 | .4943 | 44 | .5057 | 40′ |
| 30′ | .4781 | 40 | .9794 | 4 | .4987 | 44 | .5013 | 30′ |
| 40′ | .4821 | 40 | .9790 | 4 | .5031 | 44 | .4969 | 20′ |
| 50′ | .4861 | 39 | .9786 | 4 | .5075 | 43 | .4925 | 10′ |
| 18° 0′ | 9.4900 | 39 | 9.9782 | 4 | 9.5118 | 43 | 0.4882 | 72° 0′ |
| 10′ | .4939 | 38 | .9778 | 4 | .5161 | 42 | .4839 | 50′ |
| 20′ | .4977 | 38 | .9774 | 4 | .5203 | 42 | .4797 | 40′ |
| 30′ | .5015 | 37 | .9770 | 5 | .5245 | 42 | .4755 | 30′ |
| 40′ | .5052 | 38 | .9765 | 4 | .5287 | 42 | .4713 | 20′ |
| 50′ | .5090 | 36 | .9761 | 4 | .5329 | 41 | .4671 | 10′ |
| 19° 0′ | 9.5126 | 37 | 9.9757 | 5 | 9.5370 | 41 | 0.4630 | 71° 0′ |
| 10′ | .5163 | 36 | .9752 | 4 | .5411 | 40 | .4589 | 50′ |
| 20′ | .5199 | 36 | .9748 | 5 | .5451 | 40 | .4549 | 40′ |
| 30′ | .5235 | 35 | .9743 | 4 | .5491 | 40 | .4509 | 30′ |
| 40′ | .5270 | 36 | .9739 | 5 | .5531 | 40 | .4469 | 20′ |
| 50′ | .5306 | 35 | .9734 | 4 | .5571 | 40 | .4429 | 10′ |
| 20° 0′ | 9.5341 | 34 | 9.9730 | 5 | 9.5611 | 39 | 0.4389 | 70° 0′ |
| 10′ | .5375 | 34 | .9725 | 4 | .5650 | 39 | .4350 | 50′ |
| 20′ | .5409 | 34 | .9721 | 5 | .5689 | 38 | .4311 | 40′ |
| 30′ | .5443 | 34 | .9716 | 5 | .5727 | 39 | .4273 | 30′ |
| 40′ | .5477 | 33 | .9711 | 5 | .5766 | 38 | .4234 | 20′ |
| 50′ | .5510 | 33 | .9706 | 4 | .5804 | 38 | .4196 | 10′ |
| 21° 0′ | 9.5543 | 33 | 9.9702 | 5 | 9.5842 | 37 | 0.4158 | 69° 0′ |
| 10′ | .5576 | 33 | .9697 | 5 | .5879 | 38 | .4121 | 50′ |
| 20′ | .5609 | 32 | .9692 | 5 | .5917 | 37 | .4083 | 40′ |
| 30′ | .5641 | 32 | .9687 | 5 | .5954 | 37 | .4046 | 30′ |
| 40′ | .5673 | 31 | .9682 | 5 | .5991 | 37 | .4009 | 20′ |
| 50′ | .5704 | 32 | .9677 | 5 | .6028 | 36 | .3972 | 10′ |
| 22° 0′ | 9.5736 | 31 | 9.9672 | 5 | 9.6064 | 36 | 0.3936 | 68° 0′ |
| 10′ | .5767 | 31 | .9667 | 6 | .6100 | 36 | .3900 | 50′ |
| 20′ | .5798 | 30 | .9661 | 5 | .6136 | 36 | .3864 | 40′ |
| 30′ | .5828 | | .9656 | | .6172 | | .3828 | 30′ |
| | log cos | d | log sin | d | log cot | d | log tan | $x$ |

**Prop. Parts**

| | 50 | 49 | 48 | 47 |
|---|---|---|---|---|
| 1 | 5.0 | 4.9 | 4.8 | 4.7 |
| 2 | 10.0 | 9.8 | 9.6 | 9.4 |
| 3 | 15.0 | 14.7 | 14.4 | 14.1 |
| 4 | 20.0 | 19.6 | 19.2 | 18.8 |
| 5 | 25.0 | 24.5 | 24.0 | 23.5 |
| 6 | 30.0 | 29.4 | 28.8 | 28.2 |
| 7 | 35.0 | 34.3 | 33.6 | 32.9 |
| 8 | 40.0 | 39.2 | 38.4 | 37.6 |
| 9 | 45.0 | 44.1 | 43.2 | 42.3 |

| | 46 | 45 | 44 | 43 |
|---|---|---|---|---|
| 1 | 4.6 | 4.5 | 4.4 | 4.3 |
| 2 | 9.2 | 9.0 | 8.8 | 8.6 |
| 3 | 13.8 | 13.5 | 13.2 | 12.9 |
| 4 | 18.4 | 18.0 | 17.6 | 17.2 |
| 5 | 23.0 | 22.5 | 22.0 | 21.5 |
| 6 | 27.6 | 27.0 | 26.4 | 25.8 |
| 7 | 32.2 | 31.5 | 30.8 | 30.1 |
| 8 | 36.8 | 36.0 | 35.2 | 34.4 |
| 9 | 41.4 | 40.5 | 39.6 | 38.7 |

| | 42 | 41 | 40 | 39 |
|---|---|---|---|---|
| 1 | 4.2 | 4.1 | 4.0 | 3.9 |
| 2 | 8.4 | 8.2 | 8.0 | 7.8 |
| 3 | 12.6 | 12.3 | 12.0 | 11.7 |
| 4 | 16.8 | 16.4 | 16.0 | 15.6 |
| 5 | 21.0 | 20.5 | 20.0 | 19.5 |
| 6 | 25.2 | 24.6 | 24.0 | 23.4 |
| 7 | 29.4 | 28.7 | 28.0 | 27.3 |
| 8 | 33.6 | 32.8 | 32.0 | 31.2 |
| 9 | 37.8 | 36.9 | 36.0 | 35.1 |

| | 38 | 37 | 36 | 35 |
|---|---|---|---|---|
| 1 | 3.8 | 3.7 | 3.6 | 3.5 |
| 2 | 7.6 | 7.4 | 7.2 | 7.0 |
| 3 | 11.4 | 11.1 | 10.8 | 10.5 |
| 4 | 15.2 | 14.8 | 14.4 | 14.0 |
| 5 | 19.0 | 18.5 | 18.0 | 17.5 |
| 6 | 22.8 | 22.2 | 21.6 | 21.0 |
| 7 | 26.6 | 25.9 | 25.2 | 24.5 |
| 8 | 30.4 | 29.6 | 28.8 | 28.0 |
| 9 | 34.2 | 33.3 | 32.4 | 31.5 |

| | 34 | 33 | 32 | 31 |
|---|---|---|---|---|
| 1 | 3.4 | 3.3 | 3.2 | 3.1 |
| 2 | 6.8 | 6.6 | 6.4 | 6.2 |
| 3 | 10.2 | 9.9 | 9.6 | 9.3 |
| 4 | 13.6 | 13.2 | 12.8 | 12.4 |
| 5 | 17.0 | 16.5 | 16.0 | 15.5 |
| 6 | 20.4 | 19.8 | 19.2 | 18.6 |
| 7 | 23.8 | 23.1 | 22.4 | 21.7 |
| 8 | 27.2 | 26.4 | 25.6 | 24.8 |
| 9 | 30.6 | 29.7 | 28.8 | 27.9 |

**Prop. Parts**

## Tangents, and Cotangents. IV

| $x$ | log sin | d | log cos | d | log tan | d | log cot | |
|---|---|---|---|---|---|---|---|---|
| 30′ | 9.5828 | 31 | 9.9656 | 5 | 9.6172 | 36 | 0.3828 | 30′ |
| 40′ | .5859 | 30 | .9651 | 5 | .6208 | 35 | .3792 | 20′ |
| 50′ | .5889 | 30 | .9646 | 6 | .6243 | 36 | .3757 | 10′ |
| **23° 0′** | 9.5919 | 29 | 9.9640 | 5 | 9.6279 | 35 | 0.3721 | **67° 0′** |
| 10′ | .5948 | 30 | .9635 | 6 | .6314 | 34 | .3686 | 50′ |
| 20′ | .5978 | 29 | .9629 | 5 | .6348 | 35 | .3652 | 40′ |
| 30′ | .6007 | 29 | .9624 | 6 | .6383 | 34 | .3617 | 30′ |
| 40′ | .6036 | 29 | .9618 | 5 | .6417 | 35 | .3583 | 20′ |
| 50′ | .6065 | 28 | .9613 | 6 | .6452 | 34 | .3548 | 10′ |
| **24° 0′** | 9.6093 | 28 | 9.9607 | 5 | 9.6486 | 34 | 0.3514 | **66° 0′** |
| 10′ | .6121 | 28 | .9602 | 6 | .6520 | 33 | .3480 | 50′ |
| 20′ | .6149 | 28 | .9596 | 6 | .6553 | 34 | .3447 | 40′ |
| 30′ | .6177 | 28 | .9590 | 6 | .6587 | 33 | .3413 | 30′ |
| 40′ | .6205 | 27 | .9584 | 5 | .6620 | 34 | .3380 | 20′ |
| 50′ | .6232 | 27 | .9579 | 6 | .6654 | 33 | .3346 | 10′ |
| **25° 0′** | 9.6259 | 27 | 9.9573 | 6 | 9.6687 | 33 | 0.3313 | **65° 0′** |
| 10′ | .6286 | 27 | .9567 | 6 | .6720 | 32 | .3280 | 50′ |
| 20′ | .6313 | 27 | .9561 | 6 | .6752 | 33 | .3248 | 40′ |
| 30′ | .6340 | 26 | .9555 | 6 | .6785 | 32 | .3215 | 30′ |
| 40′ | .6366 | 26 | .9549 | 6 | .6817 | 33 | .3183 | 20′ |
| 50′ | .6392 | 26 | .9543 | 6 | .6850 | 32 | .3150 | 10′ |
| **26° 0′** | 9.6418 | 26 | 9.9537 | 7 | 9.6882 | 32 | 0.3118 | **64° 0′** |
| 10′ | .6444 | 26 | .9530 | 6 | .6914 | 32 | .3086 | 50′ |
| 20′ | .6470 | 25 | .9524 | 6 | .6946 | 31 | .3054 | 40′ |
| 30′ | .6495 | 26 | .9518 | 6 | .6977 | 32 | .3023 | 30′ |
| 40′ | .6521 | 25 | .9512 | 7 | .7009 | 31 | .2991 | 20′ |
| 50′ | .6546 | 24 | .9505 | 6 | .7040 | 32 | .2960 | 10′ |
| **27° 0′** | 9.6570 | 25 | 9.9499 | 7 | 9.7072 | 31 | 0.2928 | **63° 0′** |
| 10′ | .6595 | 25 | .9492 | 6 | .7103 | 31 | .2897 | 50′ |
| 20′ | .6620 | 24 | .9486 | 7 | .7134 | 31 | .2866 | 40′ |
| 30′ | .6644 | 24 | .9479 | 6 | .7165 | 31 | .2835 | 30′ |
| 40′ | .6668 | 24 | .9473 | 7 | .7196 | 30 | .2804 | 20′ |
| 50′ | .6692 | 24 | .9466 | 7 | .7226 | 31 | .2774 | 10′ |
| **28° 0′** | 9.6716 | 24 | 9.9459 | 6 | 9.7257 | 30 | 0.2743 | **62° 0′** |
| 10′ | .6740 | 23 | .9453 | 7 | .7287 | 30 | .2713 | 50′ |
| 20′ | .6763 | 24 | .9446 | 7 | .7317 | 31 | .2683 | 40′ |
| 30′ | .6787 | 23 | .9439 | 7 | .7348 | 30 | .2652 | 30′ |
| 40′ | .6810 | 23 | .9432 | 7 | .7378 | 30 | .2622 | 20′ |
| 50′ | .6833 | 23 | .9425 | 7 | .7408 | 30 | .2592 | 10′ |
| **29° 0′** | 9.6856 | 22 | 9.9418 | 7 | 9.7438 | 29 | 0.2562 | **61° 0′** |
| 10′ | .6878 | 23 | .9411 | 7 | .7467 | 30 | .2533 | 50′ |
| 20′ | .6901 | 22 | .9404 | 7 | .7497 | 29 | .2503 | 40′ |
| 30′ | .6923 | 23 | .9397 | 7 | .7526 | 30 | .2474 | 30′ |
| 40′ | .6946 | 22 | .9390 | 7 | .7556 | 29 | .2444 | 20′ |
| 50′ | .6968 | 22 | .9383 | 8 | .7585 | 29 | .2415 | 10′ |
| **30° 0′** | 9.6990 | | 9.9375 | | 9.7614 | | 0.2386 | **60° 0′** |
| | log cos | d | log sin | d | log cot | d | log tan | $x$ |

**Prop. Parts**

| | 36 | 35 | 34 | 33 | 32 | 31 | 30 | 29 | 28 | 27 | 26 | 25 | 24 | 23 | 22 |
|---|---|---|---|---|---|---|---|---|---|---|---|---|---|---|---|
| 1 | 3.6 | 3.5 | 3.4 | 3.3 | 3.2 | 3.1 | 3.0 | 2.9 | 2.8 | 2.7 | 2.6 | 2.5 | 2.4 | 2.3 | 2.2 |
| 2 | 7.2 | 7.0 | 6.8 | 6.6 | 6.4 | 6.2 | 6.0 | 5.8 | 5.6 | 5.4 | 5.2 | 5.0 | 4.8 | 4.6 | 4.4 |
| 3 | 10.8 | 10.5 | 10.2 | 9.9 | 9.6 | 9.3 | 9.0 | 8.7 | 8.4 | 8.1 | 7.8 | 7.5 | 7.2 | 6.9 | 6.6 |
| 4 | 14.4 | 14.0 | 13.6 | 13.2 | 12.8 | 12.4 | 12.0 | 11.6 | 11.2 | 10.8 | 10.4 | 10.0 | 9.6 | 9.2 | 8.8 |
| 5 | 18.0 | 17.5 | 17.0 | 16.5 | 16.0 | 15.5 | 15.0 | 14.5 | 14.0 | 13.5 | 13.0 | 12.5 | 12.0 | 11.5 | 11.0 |
| 6 | 21.6 | 21.0 | 20.4 | 19.8 | 19.2 | 18.6 | 18.0 | 17.4 | 16.8 | 16.2 | 15.6 | 15.0 | 14.4 | 13.8 | 13.2 |
| 7 | 25.2 | 24.5 | 23.8 | 23.1 | 22.4 | 21.7 | 21.0 | 20.3 | 19.6 | 18.9 | 18.2 | 17.5 | 16.8 | 16.1 | 15.4 |
| 8 | 28.8 | 28.0 | 27.2 | 26.4 | 25.6 | 24.8 | 24.0 | 23.2 | 22.4 | 21.6 | 20.8 | 20.0 | 19.2 | 18.4 | 17.6 |
| 9 | 32.4 | 31.5 | 30.6 | 29.7 | 28.8 | 27.9 | 27.0 | 26.1 | 25.2 | 24.3 | 23.4 | 22.5 | 21.6 | 20.7 | 19.8 |

## IV. Logarithms of Sines, Cosines,

| x | log sin | d | log cos | d | log tan | d | log cot | |
|---|---|---|---|---|---|---|---|---|
| **30° 0′** | 9.6990 | 22 | 9.9375 | 7 | 9.7614 | 30 | 0.2386 | **60° 0′** |
| 10′ | .7012 | 21 | .9368 | 7 | .7644 | 29 | .2356 | 50′ |
| 20′ | .7033 | 22 | .9361 | 8 | .7673 | 28 | .2327 | 40′ |
| 30′ | .7055 | 21 | .9353 | 7 | .7701 | 29 | .2299 | 30′ |
| 40′ | .7076 | 21 | .9346 | 8 | .7730 | 29 | .2270 | 20′ |
| 50′ | .7097 | 21 | .9338 | 7 | .7759 | 29 | .2241 | 10′ |
| **31° 0′** | 9.7118 | 21 | 9.9331 | 8 | 9.7788 | 28 | 0.2212 | **59° 0′** |
| 10′ | .7139 | 21 | .9323 | 8 | .7816 | 29 | .2184 | 50′ |
| 20′ | .7160 | 21 | .9315 | 7 | .7845 | 28 | .2155 | 40′ |
| 30′ | .7181 | 20 | .9308 | 8 | .7873 | 29 | .2127 | 30′ |
| 40′ | .7201 | 21 | .9300 | 8 | .7902 | 28 | .2098 | 20′ |
| 50′ | .7222 | 20 | .9292 | 8 | .7930 | 28 | .2070 | 10′ |
| **32° 0′** | 9.7242 | 20 | 9.9284 | 8 | 9.7958 | 28 | 0.2042 | **58° 0′** |
| 10′ | .7262 | 20 | .9276 | 8 | .7986 | 28 | .2014 | 50′ |
| 20′ | .7282 | 20 | .9268 | 8 | .8014 | 28 | .1986 | 40′ |
| 30′ | .7302 | 20 | .9260 | 8 | .8042 | 28 | .1958 | 30′ |
| 40′ | .7322 | 20 | .9252 | 8 | .8070 | 27 | .1930 | 20′ |
| 50′ | .7342 | 19 | .9244 | 8 | .8097 | 28 | .1903 | 10′ |
| **33° 0′** | 9.7361 | 19 | 9.9236 | 8 | 9.8125 | 28 | 0.1875 | **57° 0′** |
| 10′ | .7380 | 20 | .9228 | 9 | .8153 | 27 | .1847 | 50′ |
| 20′ | .7400 | 19 | .9219 | 8 | .8180 | 28 | .1820 | 40′ |
| 30′ | .7419 | 19 | .9211 | 8 | .8208 | 27 | .1792 | 30′ |
| 40′ | .7438 | 19 | .9203 | 9 | .8235 | 28 | .1765 | 20′ |
| 50′ | .7457 | 19 | .9194 | 8 | .8263 | 27 | .1737 | 10′ |
| **34° 0′** | 9.7476 | 18 | 9.9186 | 9 | 9.8290 | 27 | 0.1710 | **56° 0′** |
| 10′ | .7494 | 19 | .9177 | 8 | .8317 | 27 | .1683 | 50′ |
| 20′ | .7513 | 18 | .9169 | 9 | .8344 | 27 | .1656 | 40′ |
| 30′ | .7531 | 19 | .9160 | 9 | .8371 | 27 | .1629 | 30′ |
| 40′ | .7550 | 18 | .9151 | 9 | .8398 | 27 | .1602 | 20′ |
| 50′ | .7568 | 18 | .9142 | 8 | .8425 | 27 | .1575 | 10′ |
| **35° 0′** | 9.7586 | 18 | 9.9134 | 9 | 9.8452 | 27 | 0.1548 | **55° 0′** |
| 10′ | .7604 | 18 | .9125 | 9 | .8479 | 27 | .1521 | 50′ |
| 20′ | .7622 | 18 | .9116 | 9 | .8506 | 27 | .1494 | 40′ |
| 30′ | .7640 | 17 | .9107 | 9 | .8533 | 26 | .1467 | 30′ |
| 40′ | .7657 | 18 | .9098 | 9 | .8559 | 27 | .1441 | 20′ |
| 50′ | .7675 | 17 | .9089 | 9 | .8586 | 27 | .1414 | 10′ |
| **36° 0′** | 9.7692 | 18 | 9.9080 | 10 | 9.8613 | 26 | 0.1387 | **54° 0′** |
| 10′ | .7710 | 17 | .9070 | 9 | .8639 | 27 | .1361 | 50′ |
| 20′ | .7727 | 17 | .9061 | 9 | .8666 | 26 | .1334 | 40′ |
| 30′ | .7744 | 17 | .9052 | 10 | .8692 | 26 | .1308 | 30′ |
| 40′ | .7761 | 17 | .9042 | 9 | .8718 | 27 | .1282 | 20′ |
| 50′ | .7778 | 17 | .9033 | 10 | .8745 | 26 | .1255 | 10′ |
| **37° 0′** | 9.7795 | 16 | 9.9023 | 9 | 9.8771 | 26 | 0.1229 | **53° 0′** |
| 10′ | .7811 | 17 | .9014 | 10 | .8797 | 27 | .1203 | 50′ |
| 20′ | .7828 | 16 | .9004 | 9 | .8824 | 26 | .1176 | 40′ |
| 30′ | .7844 | | .8995 | | .8850 | | .1150 | 30′ |
| | log cos | d | log sin | d | log cot | d | log tan | x |

**Prop. Parts**

| | 30 | 29 | 28 |
|---|---|---|---|
| 1 | 3.0 | 2.9 | 2.8 |
| 2 | 6.0 | 5.8 | 5.6 |
| 3 | 9.0 | 8.7 | 8.4 |
| 4 | 12.0 | 11.6 | 11.2 |
| 5 | 15.0 | 14.5 | 14.0 |
| 6 | 18.0 | 17.4 | 16.8 |
| 7 | 21.0 | 20.3 | 19.6 |
| 8 | 24.0 | 23.2 | 22.4 |
| 9 | 27.0 | 26.1 | 25.2 |

| | 27 | 26 | 22 |
|---|---|---|---|
| 1 | 2.7 | 2.6 | 2.2 |
| 2 | 5.4 | 5.2 | 4.4 |
| 3 | 8.1 | 7.8 | 6.6 |
| 4 | 10.8 | 10.4 | 8.8 |
| 5 | 13.5 | 13.0 | 11.0 |
| 6 | 16.2 | 15.6 | 13.2 |
| 7 | 18.9 | 18.2 | 15 4 |
| 8 | 21.6 | 20.8 | 17 6 |
| 9 | 24.3 | 23.4 | 19.8 |

| | 21 | 20 | 19 |
|---|---|---|---|
| 1 | 2.1 | 2.0 | 1.9 |
| 2 | 4.2 | 4.0 | 3.8 |
| 3 | 6.3 | 6.0 | 5.7 |
| 4 | 8.4 | 8.0 | 7.6 |
| 5 | 10.5 | 10.0 | 9.5 |
| 6 | 12.6 | 12.0 | 11.4 |
| 7 | 14.7 | 14.0 | 13.3 |
| 8 | 16.8 | 16.0 | 15.2 |
| 9 | 18.9 | 18.0 | 17.1 |

| | 18 | 17 | 16 |
|---|---|---|---|
| 1 | 1.8 | 1.7 | 1.6 |
| 2 | 3.6 | 3.4 | 3.2 |
| 3 | 5.4 | 5.1 | 4.8 |
| 4 | 7.2 | 6.8 | 6.4 |
| 5 | 9.0 | 8.5 | 8.0 |
| 6 | 10.8 | 10.2 | 9.6 |
| 7 | 12.6 | 11.9 | 11.2 |
| 8 | 14.4 | 13.6 | 12.8 |
| 9 | 16.2 | 15.3 | 14.4 |

| | 9 | 8 | 7 |
|---|---|---|---|
| 1 | .9 | .8 | .7 |
| 2 | 1.8 | 1.6 | 1.4 |
| 3 | 2.7 | 2.4 | 2.1 |
| 4 | 3.6 | 3.2 | 2.8 |
| 5 | 4.5 | 4.0 | 3.5 |
| 6 | 5.4 | 4.8 | 4.2 |
| 7 | 6.3 | 5.6 | 4.9 |
| 8 | 7.2 | 6.4 | 5.6 |
| 9 | 8.1 | 7.2 | 6.3 |

**Prop. Parts**

## Tangents, and Cotangents. IV

| $x$ | log sin | d | log cos | d | log tan | d | log cot | |
|---|---|---|---|---|---|---|---|---|
| 30′ | 9.7844 | 17 | 9.8995 | 10 | 9.8850 | 26 | 0.1150 | 30′ |
| 40′ | .7861 | 16 | .8985 | 10 | .8876 | 26 | .1124 | 20′ |
| 50′ | .7877 | 16 | .8975 | 10 | .8902 | 26 | .1098 | 10′ |
| **38° 0′** | 9.7893 | 17 | 9.8965 | 10 | 9.8928 | 26 | 0.1072 | **52° 0′** |
| 10′ | .7910 | 16 | .8955 | 10 | .8954 | 26 | .1046 | 50′ |
| 20′ | .7926 | 15 | .8945 | 10 | .8980 | 26 | .1020 | 40′ |
| 30′ | .7941 | 16 | .8935 | 10 | .9006 | 26 | .0994 | 30′ |
| 40′ | .7957 | 16 | .8925 | 10 | .9032 | 26 | .0968 | 20′ |
| 50′ | .7973 | 16 | .8915 | 10 | .9058 | 26 | .0942 | 10′ |
| **39° 0′** | 9.7989 | 15 | 9.8905 | 10 | 9.9084 | 26 | 0.0916 | **51° 0′** |
| 10′ | .8004 | 16 | .8895 | 11 | .9110 | 25 | .0890 | 50′ |
| 20′ | .8020 | 15 | .8884 | 10 | .9135 | 26 | .0865 | 40′ |
| 30′ | .8035 | 15 | .8874 | 10 | .9161 | 26 | .0839 | 30′ |
| 40′ | .8050 | 16 | .8864 | 11 | .9187 | 25 | .0813 | 20′ |
| 50′ | .8066 | 15 | .8853 | 10 | .9212 | 26 | .0788 | 10′ |
| **40° 0′** | 9.8081 | 15 | 9.8843 | 11 | 9.9238 | 26 | 0.0762 | **50° 0′** |
| 10′ | .8096 | 15 | .8832 | 11 | .9264 | 25 | .0736 | 50′ |
| 20′ | .8111 | 14 | .8821 | 11 | .9289 | 26 | .0711 | 40′ |
| 30′ | .8125 | 15 | .8810 | 10 | .9315 | 26 | .0685 | 30′ |
| 40′ | .8140 | 15 | .8800 | 11 | .9341 | 25 | .0659 | 20′ |
| 50′ | .8155 | 14 | .8789 | 11 | .9366 | 26 | .0634 | 10′ |
| **41° 0′** | 9.8169 | 15 | 9.8778 | 11 | 9.9392 | 25 | 0.0608 | **49° 0′** |
| 10′ | .8184 | 14 | .8767 | 11 | .9417 | 26 | .0583 | 50′ |
| 20′ | .8198 | 15 | .8756 | 11 | .9443 | 25 | .0557 | 40′ |
| 30′ | .8213 | 14 | .8745 | 12 | .9468 | 26 | .0532 | 30′ |
| 40′ | .8227 | 14 | .8733 | 11 | .9494 | 25 | .0506 | 20′ |
| 50′ | .8241 | 14 | .8722 | 11 | .9519 | 25 | .0481 | 10′ |
| **42° 0′** | 9.8255 | 14 | 9.8711 | 12 | 9.9544 | 26 | 0.0456 | **48° 0′** |
| 10′ | .8269 | 14 | .8699 | 11 | .9570 | 25 | .0430 | 50′ |
| 20′ | .8283 | 14 | .8688 | 12 | .9595 | 26 | .0405 | 40′ |
| 30′ | .8297 | 14 | .8676 | 11 | .9621 | 25 | .0379 | 30′ |
| 40′ | .8311 | 13 | .8665 | 12 | .9646 | 25 | .0354 | 20′ |
| 50′ | .8324 | 14 | .8653 | 12 | .9671 | 26 | .0329 | 10′ |
| **43° 0′** | 9.8338 | 13 | 9.8641 | 12 | 9.9697 | 25 | 0.0303 | **47° 0′** |
| 10′ | .8351 | 14 | .8629 | 11 | .9722 | 25 | .0278 | 50′ |
| 20′ | .8365 | 13 | .8618 | 12 | .9747 | 25 | .0253 | 40′ |
| 30′ | .8378 | 13 | .8606 | 12 | .9772 | 26 | .0228 | 30′ |
| 40′ | .8391 | 14 | .8594 | 12 | .9798 | 25 | .0202 | 20′ |
| 50′ | .8405 | 13 | .8582 | 13 | .9823 | 25 | .0177 | 10′ |
| **44° 0′** | 9.8418 | 13 | 9.8569 | 12 | 9.9848 | 26 | 0.0152 | **46° 0′** |
| 10′ | .8431 | 13 | .8557 | 12 | .9874 | 25 | .0126 | 50′ |
| 20′ | .8444 | 13 | .8545 | 13 | .9899 | 25 | .0101 | 40′ |
| 30′ | .8457 | 12 | .8532 | 12 | .9924 | 25 | .0076 | 30′ |
| 40′ | .8469 | 13 | .8520 | 13 | .9949 | 26 | .0051 | 20′ |
| 50′ | .8482 | 13 | .8507 | 12 | .9975 | 25 | .0025 | 10′ |
| **45° 0′** | 9.8495 | | 9.8495 | | 0.0000 | | 0.0000 | **45° 0′** |
| | log cos | d | log sin | d | log cot | d | log tan | $x$ |

**Prop. Parts**

| | 26 | 25 |
|---|---|---|
| 1 | 2.6 | 2.5 |
| 2 | 5.2 | 5.0 |
| 3 | 7.8 | 7.5 |
| 4 | 10.4 | 10.0 |
| 5 | 13.0 | 12.5 |
| 6 | 15.6 | 15.0 |
| 7 | 18.2 | 17.5 |
| 8 | 20.8 | 20.0 |
| 9 | 23.4 | 22.5 |

| | 17 | 16 | 15 |
|---|---|---|---|
| 1 | 1.7 | 1.6 | 1.5 |
| 2 | 3.4 | 3.2 | 3.0 |
| 3 | 5.1 | 4.8 | 4.5 |
| 4 | 6.8 | 6.4 | 6.0 |
| 5 | 8.5 | 8.0 | 7.5 |
| 6 | 10.2 | 9.6 | 9.0 |
| 7 | 11.9 | 11.2 | 10.5 |
| 8 | 13.6 | 12.8 | 12.0 |
| 9 | 15.3 | 14.4 | 13.5 |

| | 14 | 13 | 12 |
|---|---|---|---|
| 1 | 1.4 | 1.3 | 1.2 |
| 2 | 2.8 | 2.6 | 2.4 |
| 3 | 4.2 | 3.9 | 3.6 |
| 4 | 5.6 | 5.2 | 4.8 |
| 5 | 7.0 | 6.5 | 6.0 |
| 6 | 8.4 | 7.8 | 7.2 |
| 7 | 9.8 | 9.1 | 8.4 |
| 8 | 11.2 | 10.4 | 9.6 |
| 9 | 12.6 | 11.7 | 10.8 |

| | 11 | 10 |
|---|---|---|
| 1 | 1.1 | 1.0 |
| 2 | 2.2 | 2.0 |
| 3 | 3.3 | 3.0 |
| 4 | 4.4 | 4.0 |
| 5 | 5.5 | 5.0 |
| 6 | 6.6 | 6.0 |
| 7 | 7.7 | 7.0 |
| 8 | 8.8 | 8.0 |
| 9 | 9.9 | 9.0 |

**Prop. Parts**

## V. Natural Functions, with Radian Measure

| $x$ Rdn. | Degrees in $x$ | sin $x$ | cos $x$ | tan $x$ | $x$ Rdn. | Degrees in $x$ | sin $x$ | cos $x$ | tan $x$ |
|---|---|---|---|---|---|---|---|---|---|
| .00 | 0°00′ | .0000 | 1.0000 | .0000 | .40 | 22°55′ | .3894 | .9211 | .4228 |
| .01 | 0°34′ | .0100 | .9999 | .0100 | .41 | 23°30′ | .3986 | .9171 | .4346 |
| .02 | 1°09′ | .0200 | .9998 | .0200 | .42 | 24°04′ | .4078 | .9131 | .4466 |
| .03 | 1°43′ | .0300 | .9996 | .0300 | .43 | 24°38′ | .4169 | .9090 | .4586 |
| .04 | 2°18′ | .0400 | .9992 | .0400 | .44 | 25°13′ | .4259 | .9048 | .4708 |
| .05 | 2°52′ | .0500 | .9988 | .0500 | .45 | 25°47′ | .4350 | .9004 | .4831 |
| .06 | 3°26′ | .0600 | .9982 | .0601 | .46 | 26°21′ | .4440 | .8960 | .4954 |
| .07 | 4°01′ | .0699 | .9976 | .0701 | .47 | 26°56′ | .4529 | .8916 | .5080 |
| .08 | 4°35′ | .0799 | .9968 | .0802 | .48 | 27°30′ | .4618 | .8870 | .5206 |
| .09 | 5°09′ | .0899 | .9960 | .0902 | .49 | 28°04′ | .4706 | .8823 | .5334 |
| .10 | 5°44′ | .0998 | .9950 | .1003 | .50 | 28°39′ | .4794 | .8776 | .5463 |
| .11 | 6°18′ | .1098 | .9940 | .1104 | .51 | 29°13′ | .4882 | .8727 | .5594 |
| .12 | 6°52′ | .1197 | .9928 | .1206 | .52 | 29°48′ | .4969 | .8678 | .5726 |
| .13 | 7°27′ | .1296 | .9916 | .1307 | .53 | 30°22′ | .5055 | .8628 | .5859 |
| .14 | 8°01′ | .1395 | .9902 | .1409 | .54 | 30°56′ | .5141 | .8577 | .5994 |
| .15 | 8°36′ | .1494 | .9888 | .1511 | .55 | 31°31′ | .5227 | .8525 | .6131 |
| .16 | 9°10′ | .1593 | .9872 | .1614 | .56 | 32°05′ | .5312 | .8473 | .6270 |
| .17 | 9°44′ | .1692 | .9856 | .1717 | .57 | 32°40′ | .5396 | .8419 | .6410 |
| .18 | 10°19′ | .1790 | .9838 | .1820 | .58 | 33°14′ | .5480 | .8365 | .6552 |
| .19 | 10°53′ | .1889 | .9820 | .1923 | .59 | 33°48′ | .5564 | .8309 | .6696 |
| .20 | 11°28′ | .1987 | .9801 | .2027 | .60 | 34°23′ | .5646 | .8253 | .6841 |
| .21 | 12°02′ | .2085 | .9780 | .2131 | .61 | 34°57′ | .5729 | .8196 | .6989 |
| .22 | 12°36′ | .2182 | .9759 | .2236 | .62 | 35°31′ | .5810 | .8139 | .7139 |
| .23 | 13°11′ | .2280 | .9737 | .2341 | .63 | 36°06′ | .5891 | .8080 | .7291 |
| .24 | 13°45′ | .2377 | .9713 | .2447 | .64 | 36°40′ | .5972 | .8021 | .7445 |
| .25 | 14°19′ | .2474 | .9689 | .2553 | .65 | 37°14′ | .6052 | .7961 | .7602 |
| .26 | 14°54′ | .2571 | .9664 | .2660 | .66 | 37°49′ | .6131 | .7900 | .7761 |
| .27 | 15°28′ | .2667 | .9638 | .2768 | .67 | 38°23′ | .6210 | .7838 | .7922 |
| .28 | 16°03′ | .2764 | .9611 | .2876 | .68 | 38°58′ | .6288 | .7776 | .8087 |
| .29 | 16°37′ | .2860 | .9582 | .2984 | .69 | 39°32′ | .6365 | .7712 | .8253 |
| .30 | 17°11′ | .2955 | .9553 | .3093 | .70 | 40°06′ | .6442 | .7648 | .8423 |
| .31 | 17°46′ | .3051 | .9523 | .3203 | .71 | 40°41′ | .6518 | .7584 | .8595 |
| .32 | 18°20′ | .3146 | .9492 | .3314 | .72 | 41°15′ | .6594 | .7518 | .8771 |
| .33 | 18°54′ | .3240 | .9460 | .3425 | .73 | 41°50′ | .6669 | .7452 | .8949 |
| .34 | 19°29′ | .3335 | .9428 | .3537 | .74 | 42°24′ | .6743 | .7385 | .9131 |
| .35 | 20°03′ | .3429 | .9394 | .3650 | .75 | 42°58′ | .6816 | .7317 | .9316 |
| .36 | 20°38′ | .3523 | .9359 | .3764 | .76 | 43°33′ | .6889 | .7248 | .9504 |
| .37 | 21°12′ | .3616 | .9323 | .3879 | .77 | 44°07′ | .6961 | .7179 | .9697 |
| .38 | 21°46′ | .3709 | .9287 | .3994 | .78 | 44°41′ | .7033 | .7109 | .9893 |
| .39 | 22°21′ | .3802 | .9249 | .4110 | .79 | 45°16′ | .7104 | .7038 | 1.0092 |
| $x$ Rdn. | Degrees in $x$ | sin $x$ | cos $x$ | tan $x$ | $x$ Rdn. | Degrees in $x$ | sin $x$ | cos $x$ | tan $x$ |

## V. Natural Functions, with Radian Measure

| x Rdn. | Degrees in x | sin x | cos x | tan x | x Rdn. | Degrees in x | sin x | cos x | tan x |
|---|---|---|---|---|---|---|---|---|---|
| .80 | 45°50′ | .7174 | .6967 | 1.0296 | 1.20 | 68°45′ | .9320 | .3624 | 2.5722 |
| .81 | 46°25′ | .7243 | .6895 | 1.0505 | 1.21 | 69°20′ | .9356 | .3530 | 2.6503 |
| .82 | 46°59′ | .7312 | .6822 | 1.0717 | 1.22 | 69°54′ | .9391 | .3436 | 2.7328 |
| .83 | 47°33′ | .7379 | .6749 | 1.0934 | 1.23 | 70°28′ | .9425 | .3342 | 2.8198 |
| .84 | 48°08′ | .7446 | .6675 | 1.1156 | 1.24 | 71°03′ | .9458 | .3248 | 2.9119 |
| .85 | 48°42′ | .7513 | .6600 | 1.1383 | 1.25 | 71°37′ | .9490 | .3153 | 3.0096 |
| .86 | 49°16′ | .7578 | .6524 | 1.1616 | 1.26 | 72°12′ | .9521 | .3058 | 3.1133 |
| .87 | 49°51′ | .7643 | .6448 | 1.1853 | 1.27 | 72°46′ | .9551 | .2963 | 3.2236 |
| .88 | 50°25′ | .7707 | .6372 | 1.2097 | 1.28 | 73°20′ | .9580 | .2867 | 3.3413 |
| .89 | 51°00′ | .7771 | .6294 | 1.2346 | 1.29 | 73°55′ | .9608 | .2771 | 3.4672 |
| .90 | 51°34′ | .7833 | .6216 | 1.2602 | 1.30 | 74°29′ | .9636 | .2675 | 3.6021 |
| .91 | 52°08′ | .7895 | .6138 | 1.2864 | 1.31 | 75°03′ | .9662 | .2578 | 3.7471 |
| .92 | 52°43′ | .7956 | .6058 | 1.3133 | 1.32 | 75°38′ | .9687 | .2482 | 3.9033 |
| .93 | 53°17′ | .8016 | .5978 | 1.3409 | 1.33 | 76°12′ | .9712 | .2385 | 4.0723 |
| .94 | 53°52′ | .8076 | .5898 | 1.3692 | 1.34 | 76°47′ | .9735 | .2288 | 4.2556 |
| .95 | 54°26′ | .8134 | .5817 | 1.3984 | 1.35 | 77°21′ | .9757 | .2190 | 4.4552 |
| .96 | 55°00′ | .8192 | .5735 | 1.4284 | 1.36 | 77°55′ | .9779 | .2092 | 4.6734 |
| .97 | 55°35′ | .8249 | .5653 | 1.4592 | 1.37 | 78°30′ | .9799 | .1994 | 4.9131 |
| .98 | 56°09′ | .8305 | .5570 | 1.4910 | 1.38 | 79°04′ | .9818 | .1896 | 5.1774 |
| .99 | 56°43′ | .8360 | .5487 | 1.5237 | 1.39 | 79°38′ | .9837 | .1798 | 5.4707 |
| 1.00 | 57°18′ | .8415 | .5403 | 1.5574 | 1.40 | 80°13′ | .9854 | .1700 | 5.7979 |
| 1.01 | 57°52′ | .8468 | .5319 | 1.5922 | 1.41 | 80°47′ | .9871 | .1601 | 6.1654 |
| 1.02 | 58°26′ | .8521 | .5234 | 1.6281 | 1.42 | 81°22′ | .9886 | .1502 | 6.5811 |
| 1.03 | 59°01′ | .8573 | .5148 | 1.6652 | 1.43 | 81°56′ | .9901 | .1403 | 7.0555 |
| 1.04 | 59°35′ | .8624 | .5062 | 1.7036 | 1.44 | 82°30′ | .9915 | .1304 | 7.6018 |
| 1.05 | 60°10′ | .8674 | .4976 | 1.7433 | 1.45 | 83°05′ | .9927 | .1205 | 8.2381 |
| 1.06 | 60°44′ | .8724 | .4889 | 1.7844 | 1.46 | 83°39′ | .9939 | .1106 | 8.9886 |
| 1.07 | 61°18′ | .8772 | .4801 | 1.8270 | 1.47 | 84°14′ | .9949 | .1006 | 9.8874 |
| 1.08 | 61°53′ | .8820 | .4713 | 1.8712 | 1.48 | 84°48′ | .9959 | .0907 | 10.983 |
| 1.09 | 62°27′ | .8866 | .4625 | 1.9171 | 1.49 | 85°22′ | .9967 | .0807 | 12.350 |
| 1.10 | 63°02′ | .8912 | .4536 | 1.9648 | 1.50 | 85°57′ | .9975 | .0707 | 14.101 |
| 1.11 | 63°36′ | .8957 | .4447 | 2.0143 | 1.51 | 86°31′ | .9982 | .0608 | 16.428 |
| 1.12 | 64°10′ | .9001 | .4357 | 2.0660 | 1.52 | 87°05′ | .9987 | .0508 | 19.670 |
| 1.13 | 64°45′ | .9044 | .4267 | 2.1198 | 1.53 | 87°40′ | .9992 | .0408 | 24.498 |
| 1.14 | 65°19′ | .9086 | .4176 | 2.1759 | 1.54 | 88°14′ | .9995 | .0308 | 32.461 |
| 1.15 | 65°53′ | .9128 | .4085 | 2.2345 | 1.55 | 88°48′ | .9998 | .0208 | 48.078 |
| 1.16 | 66°28′ | .9168 | .3993 | 2.2958 | 1.56 | 89°23′ | .9999 | .0108 | 92.620 |
| 1.17 | 67°02′ | .9208 | .3902 | 2.3600 | 1.57 | 89°57′ | 1.0000 | .0008 | 1255.8 |
| 1.18 | 67°36′ | .9246 | .3809 | 2.4273 | 1.58 | 90°32′ | 1.0000 | −.0092 | −108.65 |
| 1.19 | 68°11′ | .9284 | .3717 | 2.4979 | 1.59 | 91°06′ | .9998 | −.0192 | −52.067 |
| x Rdn. | Degrees in x | sin x | cos x | tan x | x Rdn. | Degrees in x | sin x | cos x | tan x |

# ANSWERS TO ODD-NUMBERED EXERCISES

**Pages 2–3**

**1.** (a) $=$; (c) $\in$; (e) $\notin$. **2.** (a) 5, 8, 11, 14, 17; (c) 2, 4, 8, 16, 32. **3.** The set of all integers, the empty set. **5.** Yes. Every element of the set on the left if also an element of the set on the right. **6.** (a) true; (c) false; (e) false; (g) true.

**Pages 7–8**

**1.** $PQ$ and $QP$ are numbers, the others, sets.

**Pages 12–13**

**1.** Approx. 108 ft. **3.** (a) $\frac{OB}{OB'}$ or $\frac{OA}{OA'}$; (c) $\frac{A'B'}{OA'}$. **5.** $2x/7$. **7.** 4.4 in.

**Pages 15–17**

**1.** 2; 2. **5.** 16 ft. **7.** All except (c). **11.** Approx. 30 miles.

**Page 18**

**1.** Conclusion: the order of points is immaterial. **2.** (a) $5\sqrt{2}$; (c) 5. **7.** Circle with center (2, 3) and radius 4. Yes.

**Pages 19–20***

**4.** (a) 10; (c) $5|k|$. **5.** (c) $A \cap B = A$ and $A \cup B = B$. **11.** $\frac{\sqrt{6} + \sqrt{3}}{4}$.

**Pages 23–24**

**1.** (a) I, $-325°$; (c) I, $50°$; (e) IV, $275°$; (g) not *in* any quandrant, $180°$.
**2.** (a) $-270°$; (c) $-360°$. **3.** $-225°$, $135°$. **5.** $970°$. **7.** (a) $28°$; (c) $-340°$.
**8.** (b) $90°$.

**Pages 25–26**

**1.** (a) 2; (c) $k^2 - k$. **2.** (a) a function, domain $= \{1, 2, 3, 4\}$, range $= \{2, 4, 6, 7\}$; (c) not a function. **3.** $(-5, 0)$, $(-4, 3)$, $(-3, 4)$, $(0, 5)$, $(3, 4)$, $(4, 3)$, $(5, 0)$. **5.** Not a function; a given mother may have more than one son.

* Miscellaneous Exercises for each chapter.

**Pages 29–32**

**3.** The cosecant. **7.** (a) $-0.87$; (c) $-0.50$; (e) $-0.97$; (g) $-0.71$. **8.** (a) III; (c) IV. **9.** (a) $\sin\theta = 4/5$, $\cos\theta = -3/5$, $\tan\theta = -4/3$, $\cot\theta = -3/4$, $\sec\theta = -5/3$, $\csc\theta = 5/4$; (c) $\sin\theta = -3/5$, $\cos\theta = 4/5$, $\tan\theta = -3/4$, and so forth; (e) $\sin\theta = -15/17$, $\cos\theta = -8/17$, $\tan\theta = 15/8$, and so forth. **11.** $\tan\theta = 3/4$, $\sin\theta = -3/5$. **13.** $\tan\theta = 1$. **15.** (a) $\tan\theta > \sin\theta$ because $0 < x < r$, so that $\frac{y}{x} > \frac{y}{r}$; (c) $\sec\theta > \tan\theta$. **18.** (a) $\sin 28°$; (c) $\cot 14°$; (e) $\tan 11°35'$; (g) $\csc\theta$; (h) $\cos(90° - \theta)$, since the angle $90° - \theta$ *is* between 0° and 45°.

**Page 33**

**3.** (a) cotangent and cosecant; (c) cotangent and cosecant; (d) tangent and secant. **7.** (a) $\sin 0° = 0$, $\cos 0° = 1$, $\tan 0° = 0$, $\cot 0°$ is not defined, $\sec 0° = 1$, $\csc 0°$ is not defined; (c) $\sin 270° = -1$, $\cos 270° = 0$, $\tan 270°$ is not defined, $\cot 270° = 0$, $\sec 270°$ is not defined, $\csc 270° = -1$; (e) same as the functional values of 90° as given in the example above.

**Page 36**

**10.** (a) $\sin 150° = 1/2$, $\cos 150° = -\sqrt{3}/2$, $\tan 150° = -1/\sqrt{3}$, $\cot 150° = -\sqrt{3}$, $\sec 150° = -2/\sqrt{3}$, $\csc 150° = 2$; (c) $\sin 330° = -1/2$, $\cos 330° = \sqrt{3}/2$, $\tan 330° = -1/\sqrt{3}$, $\cot 330° = -\sqrt{3}$; $\sec 330° = 2/\sqrt{3}$, $\csc 330° = -2$. **12.** (a) 0; (c) 1; (e) $-\sqrt{3}/2$. **13.** (a) no values; (c) odd multiples of 90°; (e) odd multiples of 90°. **14.** $\sin 36° = \dfrac{\sqrt{10 - 2\sqrt{5}}}{4}$, $\cos 36° = \dfrac{\sqrt{5} + 1}{4}$.

**Page 40**

**17.** $\alpha = 23°44'$. **19.** $\beta = 46°48'$. **21.** $\theta = 61°19'$. **23.** $\alpha = 56°26'$. **25.** $\alpha = 68°3'$. **27.** $\theta = 45°48'$. **29.** $\beta = 27°20'$.

**Page 43**

**1.** (a) $\sin 60° = \sqrt{3}/2$; (c) $-\tan 60° = -\sqrt{3}$; (e) $\sec 30° = 2/\sqrt{3}$; (g) $-\sin 30° = -1/2$; (i) $-\tan 60° = -\sqrt{3}$; (k) $\sec 60° = 2$. **2.** (a) $-\cos 20°15'$; (c) $-\tan 49°23'$; (e) $-\sec 10°$; (g) $\tan 80°$; (i) $\sin 59°42'$; (k) $-\csc 81°12'$. **3.** (a) $\theta = 150°$; (c) $\theta = 135°$; (e) $\theta = 300°$; (g) $\theta = 141°35'$.

**Pages 44–45***

**1.** (a) true; (c) true; (e) false; (g) false; (i) true. **4.** (a) $\theta = 120°$; (c) $\theta = 120°$; (e) $\theta = 135°$. **7.** $\theta$ is an odd multiple of 90°. **11.** $\sin\theta = 1$ if $\theta$ is $90° + n \cdot 360°$;, where $n = 0, \pm 1, \pm 2, \pm 3, \cdots$; $\sin\theta = -1$ if $\theta = -90° + n \cdot 360°$. **13.** Tentative conclusion: $\sin(-\theta) = -\sin\theta$.

**Pages 48–49**

**1.** (a) 120°; (c) $\left(\dfrac{18}{\pi}\right)^{\circ}$ or approx. 54°4′; (e) approx. 48°8′; (g) $-270°$; (i) 225°. **2.** (a) $\pi/6$; (c) $5\pi/9$; (e) $-2\pi/3$; (g) $\pi/10$; (i) $2\pi/5$; (k) $3\pi/4$. **3.** (a) $1/2$; (c) $-1/2$; (e) $-1$. **4.** (a) $-2.1759$; (c) 0.9886; (e) 0.9757; (g) 0.8415; (i) 0.1601. **5.** (a) $\pi/3, 5\pi/3$; (c) $4\pi/3, 5\pi/3$; (e) 0.26, 3.40, both approx.

**Pages 50–51**

**1.** (a) $s = 10\pi/3$ in.; (c) $\theta = 1/2$ rdn; (e) $\theta = 4$ rdn. **3.** $4\pi$ in. **5.** 0.4 rdn. **7.** Approx. 1150 ft.

**Page 52**

**1.** 0.75 approx. rps. **3.** $\frac{5}{2}$.

**Pages 53–55***

**2.** (a) 45°; (c) 540°; (e) 240°. **4.** (a) $2\pi/5$; (c) $\pi/180$; (e) $11\pi/6$. **8.** (a) 32 sq ft; (c) 75 sq units. **11.** Approx. 1050 miles.

**Page 63**

**1.** The zeros of $\cos\theta$ are $(2n+1)\pi/2$; the zeros of $-4\cos\pi x$ are $(2n+1)\frac{1}{2}$.
**3.** (a) period $2\pi$, amplitude 2; (c) per. 4, amp. 3; (e) per. $4\pi$, amp. $\frac{1}{2}$.

**Page 67**

**1.** (a) $n\pi$; (c) no zeros; (e) $-\frac{7}{2}$.

**Page 72**

**6.** Max $\theta = 0.31 \approx \sin 0.31$; approx. 18°.

**Page 77**

**7.** (a) $\rho = 4$; (c) $\rho = 4\csc\theta$; (e) $\rho = 0$. **8.** (a) circle with center at the pole and radius 8; (c) the pole; (e) a double spiral.

**Pages 77–79***

**1.** (a) IV; (c) the cotangent; (e) the sine and the cotangent. **2.** (a) 3; (c) $\frac{1}{2}$.
**3.** (a) $-3$; (c) $-\frac{1}{2}$. **7.** $0.4\pi$ and $1.6\pi$. **11.** Intersection points: $(9, \pi/6)$, $(9, 5\pi/6)$.
**13.** $k = \frac{4}{3}$. **15.** The image of the cosine function is the interval $[-1, 1]$; the image of the tangent function and of the cotangent function is the set of all real numbers; the image of the secant function and of the cosecant function is the set $\{x \mid x \text{ is a number and } |x| \geqq 1\}$.

**Pages 83–84**

**1.** (a) $b \approx 70.74$, $c \approx 165.8$; (c) $\alpha \approx 43°33'$, $a \approx 11.84$; (e) $a = 12\sqrt{3}$, $c = 24$, $\beta = 30°$; (g) $b \approx 15.76$, $c \approx 22.61$; (i) $a \approx 6.49$; **3.** $h \approx 28.3$ ft. **5.** Approx. 5°20′.
**7.** Change the word "sine" in Ex. 6 to "cosine."

**Page 87**

**1.** (a) $b \approx 88.38$; (c) $c \approx 44.83$; (e) $\alpha \approx 52°32'$. **3.** $f \approx 3482$. **5.** $(19.89, -20.97)$.

**Page 91**

**1.** $\theta \approx 71°20'$. **3.** $d \approx 81$ ft. **5.** Approx. 447 miles, 296°30′. **7.** Approx. 80 ft.
**9.** $k = \frac{1}{2}ab\sin C$. **11.** Approx. 9°.

**Pages 92–94 (Misc. Exs. for Chap. 5)**

**1.** $\beta = 90° - \alpha$, $a = b\tan\alpha$, $c = b\sec\alpha$. **3.** Approx. 16.9 knots. **5.** $f \approx 54$ lb.
**7.** Approx. 44 sq units. **9.** Approx. $2000\sqrt{3}$ miles. **11.** $v_x \approx 3270$ fps, $v_y \approx 1250$ fps.
**13.** $AB \approx 400$ ft. **15.** $s \approx 22$ units. **17.** Approx. $\frac{1}{5}$.

**Pages 97–98**

**1.** Identity (algebraically equal). **3.** Identity (algebraically equal). **5.** Not identity (for example, not equal for $x = 3$, $y = 2$). **7.** Identity (both being positive square roots of $x^2 - 2xy + y^2$). **9.** Not identity (for example, not true for $x = 1$, $b = 1$). **11.** Not defined for $x = 5$, $x = -4$. **13.** Not defined for $x = 0$. **15.** Not defined for $x = 0$.

**Pages 102–103**

**33.** IV. **35.** II. **49.** If $n = 0, \pm 1, \pm 2, \pm 3, \cdots$, formulas (6.1), (6.3), (6.5), (6.8) are meaningless for $\theta = n\pi$; formulas (6.2), (6.3), (6.4), (6.7) are meaningless for $\theta = \frac{\pi}{2} + n\pi$.

**Page 104**

**1.** (a) $\cos 60° = \frac{1}{2}$; (c) $-\tan 150° = \tan 30° = 1/\sqrt{3}$;
(e) $\sec 225° = -\sec 45° = -\sqrt{2}$; (g) $\cos 120° = -\cos 60° = -\frac{1}{2}$.
**2.** $\cot(-\theta) = -\cot\theta$, $\sec(-\theta) = \sec\theta$, and $\csc(-\theta) = -\csc\theta$. **4.** (a) odd; (c) neither odd nor even; (e) odd; (g) even; (i) neither odd nor even.

**Page 108**

**3.** $\cos 6x$. **5.** 0. **9.** (a) $-\sqrt{2}\sin 15°$; (c) $2\sin x\sin y$. **10.** (a) $-2\sin 2x\sin x$; (c) $\cos 3x(2\cos 2x + 1)$ or $\cos x(1 + 2\cos 4x)$ according as one pairs the first and third terms or the second and third terms.

**Pages 110–111**

**3.** $\sqrt{3}/2$. **7.** $\sin 2x$. **9.** $2\sin 70°\cos 10°$. **11.** (a) $\frac{24}{25}$, 0; (c) $\frac{24}{25}$, 0.
**13.** (a) $2\sin 40°\cos 30° = \sqrt{3}\sin 40°$; (c) $2\sin 2x\cos x$.
**19.** $\cos x\cos y\cos z - \sin x\sin y\cos z - \sin x\cos y\sin z - \cos x\sin y\sin z$.

**Page 112**

**1.** $\dfrac{3-\sqrt{3}}{3+\sqrt{3}} = 2 - \sqrt{3}$. **3.** $-\sqrt{3}/3$. **11.** It shows that $x - y$ is an odd multiple of 90°.

**Pages 113–114**

**1.** $\cos 30° \approx 0.88$. **3.** $\cos 150° \approx -0.88$. **5.** $\frac{3}{4}$. **7.** 0.6, 0.8. **9.** $\frac{1}{8}$.
**17.** $\frac{1}{2}(1 - \cos 6x)$ and $\frac{1}{2}(1 + \cos 6x)$.

**Page 116**

**1.** $\frac{1}{2}\sqrt{2+\sqrt{3}}, \frac{1}{2}\sqrt{2-\sqrt{3}}, 2-\sqrt{3}$. **3.** $\frac{1}{2}\sqrt{2-\sqrt{2}}, \frac{1}{2}\sqrt{2+\sqrt{2}}, \sqrt{2}-1$.
**5.** $\sqrt{2}/10, 7\sqrt{2}/10, \frac{1}{7}$. **7.** (a) $\dfrac{1+\cos 6x}{2}$; (c) $\dfrac{1-\cos 4x}{2}$; (e) $\dfrac{1+\cos 200\pi}{2} = 1$.
**9.** $\sqrt{2}/10, 7\sqrt{2}/10$.

**Page 117**

**1.** (a) $2\sin 60°\cos 20° = \sqrt{3}\cos 20°$; (c) $2\cos 60°\cos 20° = \cos 20°$; (e) $2\cos 2\theta\sin\theta$.

**Pages 118–120***

**1.** $\sin 150° = \frac{1}{2}$. **3.** $\tan(\pi/4 + \theta - \theta) = 1$. **5.** $\sin(6x - 5x) = \sin x$.
**7.** $\cos[2(3x/2)] = \cos 3x$. **9.** 1. **11.** $\cos\theta$. **13.** $\tan^2 5\theta$. **15.** $1 + \cot^2\alpha = \csc^2\alpha$.
**17.** $2\cos A\cos B$. **19.** $\sin[\theta - (\theta - 30°)] = \frac{1}{2}$. **31.** The two expressions for $y$ are identically equal.

**Pages 127–128**

**1.** (a) $x = \pi/2 + n\cdot 2\pi$, where $n = 0, \pm 1, \pm 2, \pm 3, \cdots$; (c) $x = 3\pi/2 + 2n\pi$.
**3.** (a) $x = \pi/4 + n\pi$; (c) $x = 3\pi/4 + n\pi$. **4.** (a) $x = -\pi/4, \pi/4$; (c) $x = 45°, 135°$; (e) $x \approx -297°, -243°, 63°, 117°$; (g) using Table V, $x \approx 0.72, 5.56, 7.00, 11.84$; (i) $x = \pi/6 + n\pi$. **5.** (a) $x = 15°, 75°, 195°, 255°$; (c) $x = 45°, 105°, 165°, 225°, 285°, 345°$; (e) $x = 60°, 180°$; (g) $x = 30°, 150°, 270°$. **6.** (a) $\pi/2, 3\pi/2, 5\pi/2, 7\pi/2$; (c) $\pi/4, 5\pi/4, 9\pi/4, 13\pi/4$; (e) $60°, 240°; 420°, 600°$.

**Pages 134–136**

**1.** $\pi/3$. **3.** $-\pi/4$. **5.** $-\pi/6$. **7.** $\pi/3$. **9.** 0. **11.** $-\pi/4$. **13.** 170°. **15.** $\sqrt{3}/2$. **17.** $\frac{1}{5}$. **19.** 0. **21.** $(3 + 4\sqrt{3})/10$. **23.** $y/\sqrt{1 - y^2}$. **25.** $\sqrt{(1 - x^2)(1 - y^2)} - xy$. **27.** $3\pi/4$. **29.** (a) $x = \sin(y/2)$; (c) $x = 3 \sin(y/2)$. **37.** $x = \text{Arc} \tan k + n\pi$; $x = 3\pi/4, 7\pi/4, 11\pi/4, 15\pi/4$.

**Page 140**

**1.** $x = \pi/2, 7\pi/6, 11\pi/6$; in general, $x = n\pi + (-1)^n(-\pi/6)$, $x = n\pi + (-1)^n\pi/2$. **3.** $x = \pi/4, 3\pi/4, 5\pi/4, 7\pi/4$; $x = \pm\pi/4, + n\pi$. **5.** $y = \pi/2, 3\pi/2$; $y = \pi/2 + n\pi$. **7.** $x = 0, \pi/2, 2\pi/3, \pi, 4\pi/3, 3\pi/2, 2\pi$; $x = n\pi/2$, $x = 2n\pi \pm 2\pi/3$. **9.** $x = 0, 2\pi$; $x = 2n\pi$, $x = 2n\pi + (-1)^n(-\pi/3)$. **11.** $x = \pi/6, 5\pi/6$; $x = n\pi + (-1)^n\pi/6$. **13.** $x = 30°, 60°, 90°, 120°, 150°, 210°, 240°, 270°, 300°, 330°$. **15.** $x = 7°30', 37°30', 97°30', 127°30', 187°30', 217°30', 277°30', 307°30'$. **17.** $x = 0, 360°$. **19.** $x = 0, 210°, 330°, 360°$. **21.** $x = 1/\sqrt{5}$. **23.** $y = \frac{1}{6}$.

**Pages 140–142**

**1.** Identity. **3.** Not identity; $\theta = 0, \pi/2, \pi, 3\pi/2, 2\pi$. **5.** Not identity; $\theta = 0, \pi/2, \pi$, approx. 3.99, approx. 5.43. **7.** Identity. **9.** $x = 0, \pi/2, 7\pi/6, 11\pi/6, 2\pi$. **11.** Identity. **13.** (a) $\theta + \pi/2$; (c) $\theta - \pi/2$. **14.** (a) $x = 1/\sqrt{5}$; (c) $x = \frac{2}{15}(\sqrt{10} + \sqrt{5})$. **15.** (a) $\sqrt{10}/10$; (c) $5\sqrt{2}/7$. **17.** $\rho = 5$, $\theta = \text{Arc} \tan \frac{3}{4}$. **19.** $\rho = \sqrt{2}$, $\theta = \pi/4$; a point may have many sets of polar coordinates. **22.** (a) using $\sin\theta \cos\theta = \frac{1}{2} \sin 2\theta$, gives $2\theta = n\pi + (-1)^n \text{Arc} \sin 0$, from which $\theta = n\pi/2$; (b) combining the two formulas for the solution of $\sin\theta = \pm\frac{1}{2}$, gives the general result $\theta = n\pi \pm \pi/6$.

**Pages 148–149**

**1.** $C \approx 83°1'$, $c \approx 53.81$. **3.** $\gamma \approx 71°24'$, $b \approx 343.8$. **5.** $\alpha = 27°34'$, $c \approx 185.8$. **7.** No triangle. **9.** One triangle, $\gamma \approx 34°$, $\beta \approx 95°$, $b \approx 27$. **11.** Two triangles, $A \approx 87°$, $B \approx 40°$, $b \approx 6$; $A' \approx 93°$, $B' \approx 34°$, $b' \approx 6$. **12.** (a) none; (c) none; (e) one. **13.** $h \approx 78$ ft. **15.** Approx. 270°, 600 miles.

**Page 153**

**1.** $A \approx 76°57'$, $c \approx 11.68$. **3.** $C \approx 67°10'$, $A \approx 65°2'$, $b \approx 81.46$. **5.** $\beta = 90°$. **7.** Approx. 189 mph, 7°. **9.** Approx. 258 lb. **11.** $\sqrt{19}$. **13.** (a) $\sqrt{19}$; (b) $2\sqrt{39}$.

**Pages 157–158**

**1.** $A \approx 89°42'$, $B \approx 51°48'$. **3.** $A \approx 143°$, $C \approx 19°$. **5.** $C \approx 85°31'$, $A \approx 41°1'$. **7.** $B \approx 37°$. **9.** $C \approx 72°16'$. **11.** Approx. 326°40', 793 miles. **13.** Approx. 99°, 81°. **15.** Approx. 93°50', 299 mph.

**Pages 159–162***

**1.** Approx. 11.23 pm, 34.4 nautical miles. **3.** Approx. 357 mph, 316°40'. **5.** $h \approx 143$ ft. **7.** $\angle CBA \approx 44°50'$, $\angle CAB \approx 109°50'$. **9.** Approx. 67. **13.** Approx. 117°, 303°, 5.0 hr. **15.** Upstream at an angle of approx. 41°30' with the bank, approx. 13 min. **17.** Approx. 125 lb, 172°. **19.** Approx. 791 ft. **21.** Approx. 149,000 sq yd, or 30.8 acres.

**Page 166**

**1.** 8. **3.** $i$. **5.** $4 - i$. **7.** $-3 + i$. **9.** $-5 - 12i$. **11.** $40 + 20i$. **13.** $-2i$. **17.** $x = -1, y = 2$. **19.** $x = 2, y = 2$; $x = -2, y = -2$. **21.** $\pm\left(\frac{1}{\sqrt{2}} + \frac{1}{\sqrt{2}}i\right)$.

**Page 170**

**1.** $5(\cos 36°52' + i \sin 36°52')$. **3.** $13(\cos 112°37' + i \sin 112°37')$, the angle being approx. **5.** $4(\cos\pi + i \sin\pi)$. **7.** $\sqrt{7}(\cos 310°54' + i \sin 310°54')$, the angle being approx. **11.** $1 - 7i$. **13.** $-4 - 2i$. **15.** $-3 + 7i$. **17.** $1 - 2i$. **19.** $4i$. **21.** $-1 + i$. **23.** $-2 + 2i\sqrt{3}$. **25.** $-6$. **27.** 315°.

**Pages 172–173**

**3.** $-6 + 6i\sqrt{3}$. **5.** $4 - 4i\sqrt{3}$. **7.** $8 \text{ cis } (5\pi/4)$. **9.** $10 \text{ cis } 225°$. **11.** $-2i$. **13.** $-4 - 4i$. **15.** $8i$. **17.** $-i$. **19.** $81$. **21.** $-8i$. **23.** $\cos 2\theta = \cos^2\theta - \sin^2\theta$; $\sin 2\theta = 2 \sin\theta \cos\theta$.

**Page 174**

**1.** $\sqrt{3} + i$. **3.** $-5/3$. **5.** $-4/3$. **9.** $\dfrac{\sqrt{3}}{16} - \dfrac{1}{16}i$.

**Page 178**

**4.** (a) $2, -1 + \sqrt{3}\,i, -1 - \sqrt{3}\,i$. **5.** $\sqrt{2} \text{ cis } 67°30'$, $\sqrt{2} \text{ cis } 247°30'$, or, combining, $\pm\sqrt{2} \text{ cis } 67°30'$. **7.** Four roots, $\pm\left(\dfrac{1}{\sqrt{2}} + \dfrac{i}{\sqrt{2}}\right)$, $\pm\left(\dfrac{1}{\sqrt{2}} - \dfrac{i}{\sqrt{2}}\right)$.

**Pages 178–179***

**1.** (a) $\sqrt{2}$, $315°$; (c) $2$, $\pi/2$. **2.** (a) $-2 - i$; (c) $-1$. **3.** (a) $\frac{1}{2} + \frac{1}{2}i$; (c) $4 - 3i$. **5.** $-1 + 5i$. **7.** Approx. $10\sqrt{82}$ lb, $83°40'$ from the horizontal. **9.** $2 \text{ cis } (2C + D)$. **11.** $8 + 8i\sqrt{3}$. **13.** $-8 + 8i\sqrt{3}$. **15.** $-8i$. **17.** $\text{cis } (3\pi/8 + k\pi/2)$ with $k = 0, 1, 2, 3$. **19.** $\text{cis } 11\theta$.

# INDEX